T0214848

Wireless Multimedia Sensor Networks on Reconfigurable Hardware

Li-minn Ang • Kah Phooi Seng • Li Wern Chew
Lee Seng Yeong • Wai Chong Chia

Wireless Multimedia Sensor Networks on Reconfigurable Hardware

Information Reduction Techniques

 Springer

Li-minn Ang
Kah Phooi Seng
School of Engineering
Edith Cowan University
Joondalup, Western Australia
Australia

Li Wern Chew
Intel Architecture Group
Intel Corporation
Penang, Malaysia

Lee Seng Yeong
Wai Chong Chia
Dept. of Computer Science and Networked
 Systems
Sunway University
Selangor, Malaysia

ISBN 978-3-662-51356-9   ISBN 978-3-642-38203-1 (eBook)
DOI 10.1007/978-3-642-38203-1
Springer Heidelberg New York Dordrecht London

© Springer-Verlag Berlin Heidelberg 2013
Softcover reprint of the hardcover 1st edition 2013
This work is subject to copyright. All rights are reserved by the Publisher, whether the whole or part of
the material is concerned, specifically the rights of translation, reprinting, reuse of illustrations, recitation,
broadcasting, reproduction on microfilms or in any other physical way, and transmission or information
storage and retrieval, electronic adaptation, computer software, or by similar or dissimilar methodology
now known or hereafter developed. Exempted from this legal reservation are brief excerpts in connection
with reviews or scholarly analysis or material supplied specifically for the purpose of being entered
and executed on a computer system, for exclusive use by the purchaser of the work. Duplication of
this publication or parts thereof is permitted only under the provisions of the Copyright Law of the
Publisher's location, in its current version, and permission for use must always be obtained from Springer.
Permissions for use may be obtained through RightsLink at the Copyright Clearance Center. Violations
are liable to prosecution under the respective Copyright Law.
The use of general descriptive names, registered names, trademarks, service marks, etc. in this publication
does not imply, even in the absence of a specific statement, that such names are exempt from the relevant
protective laws and regulations and therefore free for general use.
While the advice and information in this book are believed to be true and accurate at the date of
publication, neither the authors nor the editors nor the publisher can accept any legal responsibility for
any errors or omissions that may be made. The publisher makes no warranty, express or implied, with
respect to the material contained herein.

Printed on acid-free paper

Springer is part of Springer Science+Business Media (www.springer.com)

To Grace Ang Yi-en, our blessed daughter for all the joy you bring.
  —Li-minn Ang and Kah Phooi Seng

To my parents for your unconditional love and never ending support.
  —Li Wern Chew

To my parents and loved ones for your unceasing support and love.
  —Lee Seng Yeong

To my parents, thank you for all your unconditional support.
  —Wai Chong Chia

Acknowledgements

We would like to express our deepest appreciation to all the people who contributed to the work presented in this book. We would like to thank our former Ph.D. student Christopher Ngau Wing Hong, in particular for his help on some of the work described in Chaps. 4 and 5.

Li-minn Ang and Kah Phooi Seng gratefully acknowledge support from the Centre for Communications Engineering Research at Edith Cowan University, Australia. Lee Seng Yeong and Wai Chong Chia gratefully acknowledge support from the School of Computer and Networked Systems at Sunway University, Malaysia. Li Wern Chew gratefully acknowledges support from the Intel Architecture Group at Intel, Malaysia. We gratefully acknowledge our Springer editor Ronan Nugent and the publication team for their diligent efforts and support towards the publication of this book. The support of the Malaysian eScience Fund research grants (01-02-12-SF0054, 01-02-16-SF0026) is gratefully acknowledged.

Contents

List of Figures

List of Tables

Acronyms

AIM	Attention based on information maximization
CS	Center surround
DWT	Discrete wavelet transform
EX	Instruction execution
EXMEM	Instruction execution and memory access
FOV	Field of view
FPGA	Field-programmable gate array
ID	Instruction decode
IF	Instruction fetch
MCH	Multimedia cluster head
MEM	Instruction memory access
MIPS	Microprocessor without interlocked pipeline stages
MSB	Most-significant bit
MSF	Medium spatial frequencies
NCC	Normalized cross-correlation
RAM	Random-access memory
RANSAC	Random sample consensus
SAD	Sum of absolute difference
SIFT	Shift invariant feature transform
SIG	Significant
SOT	Spatial orientation tree
SPIHT	Set partitioning in hierarchical trees
SSD	Sum of square difference
SW	Slepian–Wolf
VA	Visual attention
WB	Instruction write back
WBSME	Wavelet-based saliency map estimator
WMN	Wireless multimedia node
WMSN	Wireless multimedia sensor network
WSN	Wireless sensor network
WZ	Wyner–Ziv

Chapter 1
Introduction

Abstract This book describes how reconfigurable hardware technology like field programmable gate arrays can be used to offer a cost-efficient and flexible platform for implementation in wireless multimedia sensor networks (WMSNs). A main focus of the book is towards the development of efficient algorithms and architectures for information reduction techniques such as event detection, event compression and multi-camera processing for hardware implementation in WMSNs.

1.1 Overview

The field of wireless sensor networks (WSNs) has seen much success in diverse applications ranging from environmental monitoring, smart homes, healthcare to industrial and defence applications. Traditional WSNs capture scalar data such as temperature, vibration, pressure or humidity. Motivated by the success of WSNs and also with the emergence of new technology in the form of low-cost image sensors, researchers have proposed combining image and audio sensors with WSNs to form wireless multimedia sensor networks (WMSNs). The use of image sensors in WMSNs increases the range of potential applications because compared to scalar sensors, image sensors are able to provide more information which can be used for visual processing tasks such as detection, identification and tracking. On the one hand, WMSNs can be seen as an extension of traditional WSNs where multimedia sensors have replaced scalar sensors. On the other hand, the use of multimedia sensors in WMSNs brings with it a different set of practical and research challenges. This is because multimedia sensors, particularly image sensors, generate a high amount of data that would have to be processed and transmitted within the network. The main issue is that sensor nodes have limited battery power and hardware resources.

There are currently three options a designer has to implement algorithms in hardware: hardwired logic or application-specific integrated circuits (ASICs), software programmed microprocessors and reconfigurable computing. The

L. Ang et al., *Wireless Multimedia Sensor Networks on Reconfigurable Hardware*,
DOI 10.1007/978-3-642-38203-1_1, © Springer-Verlag Berlin Heidelberg 2013

advantage of using ASICs is that they can be designed specifically to meet the requirements for a particular application to give fast and efficient computations. The disadvantage is that any modifications to the functionality would require the entire IC to be replaced which would be a costly process. Software programmed microprocessors provide flexibility by allowing changes to the software instructions to be executed. However, this comes at a cost of performance, and, typically microprocessor-based designs are slower than ASICs due to the high overheads required for processing each instruction. The third approach is to use reconfigurable hardware which provides higher performance than software programmed microprocessors while also maintaining a higher level of flexibility than ASICs. A reconfigurable computer has the facility to make significant changes to its datapath and control. The hardware can be specially configured to perform an application-specific task such as image processing or pattern matching, and after completion the hardware can be reconfigured to perform another specific task. Currently, a popular platform for reconfigurable hardware technology is the field programmable gate array (FPGA).

The energy-efficient processing and transmission of the data within the WMSN is of primary importance to maximise the lifetime of the overall network. Depending on the application, the network should be active for a duration of time ranging from weeks to years without the need for battery replacement. To save energy in transmission, information reduction algorithms can be applied to minimise the amount of data to be transmitted. Two different visual information processing approaches can be employed to reduce the image data. The approaches can be divided into single-view approaches and multi-view approaches. Single-view approaches attempt to reduce the image data from each individual sensor node. Two techniques can be used here: event detection and event compression. The first technique uses event detection to reduce data by only transmitting frames in the network when significant events are detected. Image frames which are not significant are discarded. For example, a surveillance application could use a face event detector to decide which image frames to send to the base station. Image frames which do not contain faces do not need to be transmitted. However, the face detector would need to have low computational complexity to meet the energy requirements in the sensor node. There is a trade-off between energy required for processing and energy required for transmission. On the one hand, using an event detector in the sensor node requires more computational power. On the other hand, this could result in a saving of transmission power when frames are discarded. The other advantage of an event detector is that it could also serve as an early stage for visual pre-processing.

To perform the facial recognition process, the central computer would need to perform at least two stages. The first stage is to locate the face location in the image and the second stage would then be to perform the recognition task by comparing the facial features with a stored database. To reduce the large amount of image data for processing by the central computer, the event detector performs the face detection task, and the location(s) of the face(s) is then communicated to the central computer to perform the facial recognition task. The second technique is to perform event compression on the image frames which have to be transmitted. Event compression approaches rely on image and video compression algorithms to remove redundancy from the data. These algorithms range from current standards like JPEG, MPEG-x and H.26x to newer techniques using distributed video coding and compressive sensing. While single-view approaches attempt to reduce the scene data from each individual sensor node, multi-view approaches perform the data reduction by aggregating the data from different sensor nodes. This is possible because of the overlapping field of views (FOVs) captured by different nodes. Regions of image frames that overlap are discarded prior to transmission. However, the implementation of these information reduction algorithms would require higher computational complexity, leading to higher requirements for processing. For cost-efficient implementation in WMSNs, the algorithms would need to have low computational and memory complexity.

1.2 Book Organisation

Multimedia sensors can comprise both audio and visual sensors. For simplicity, we will use the term multimedia sensor networks even though the focus of the book is on visual sensors. The book material will be of interest to university researchers, R&D engineers, electronics, communications, and computer engineers, and graduate students working in signal and video processing, computer vision, embedded systems and sensor networks. Although some of the material presented in the book has been previously published in conference proceedings and articles [27–30, 97], most of the material has been rewritten, and algorithm descriptions have been included for practical implementations.

The remainder of the book is organised into seven chapters. Chapters 2 and 3 present background material on WMSNs and current technology. Chapter 2 discusses the emergence and architectures of WMSNs and describes the components and technology for a wireless multimedia sensor node, information processing approaches for WMSNs and various applications for WMSNs. Chapter 3 discusses the technology available and the advantages of reconfigurable hardware for WMSNs and the range of FPGA technology, internal architectures and families available. The chapter also presents an overview of programming languages for reconfigurable devices and discusses the range of available languages from lower-level languages like VHDL and Verilog to higher-level languages based on C-based languages like SystemC, SpecC and Handel-C.

Chapter 4 presents the specification for the FPGA WMSN hardware platforms. The chapter discusses the instruction set architecture and processor architecture for the datapath and control. A specification for the WMSN processor in Handel-C is included. Chapter 5 presents a case study for implementing visual event detection on the FPGA WMSN processor using visual attention. The chapter gives a background to event detection and discusses algorithms for visual saliency. The chapter also contains the description of the hardware modules and programs in assembly language. Chapter 6 presents a case study for implementing visual event compression on the FPGA WMSN processor using wavelet compression. The chapter gives a background to event compression and discusses algorithms for visual coding. The chapter also contains the description of the hardware modules and programs in assembly language. Chapter 7 presents a discussion for implementing multi-camera approaches such as visual event fusion in WMSNs. The chapter gives a background to event fusion, feature detection techniques and discusses algorithms and implementations for image stitching on the WMSN.

Chapter 2
Wireless Multimedia Sensor Network Technology

Abstract This chapter presents background material for wireless multimedia sensor network (WMSN) technology. The chapter will describe the general structure for a WMSN and various architectures and platform classifications for WMSNs. The chapter will also discuss the various components in a WMSN node such as the sensing, processing, communication, power and localisation units. The efficient processing of information in a WMSN is of primary importance, and the chapter will discuss various multi-camera network models and information reduction techniques such as event detection and event compression. The chapter concludes with a discussion of applications of WMSNs.

2.1 Introduction

The emergence of wireless multimedia sensor networks (WMSNs) is an evolutionary step for wireless sensor networks as audio and visual sensors are integrated into wireless sensor nodes. It has been a focus of research in a wide variety of areas including digital signal processing, communication, networking and control systems. WMSNs are able to store, correlate and fuse multimedia data originating from several camera input sources. The main applications for WMSNs are those that benefit from distributed and multi-camera vision systems. Deploying multiple, low-cost visual sensors both improves the coverage of the application and provides for a more robust operation. Also, multiple cameras provide redundancy to improve its reliability and usability. A single point failure will not cause a system failure, nor an obstruction or occlusion. Furthermore, multiple visual sources provide the flexibility to adaptively extract data depending on the requirements of the application. A multi-resolution description of the scene and multiple levels of abstraction can also be provided. A typical application for a WMSN would be as a surveillance and monitoring system. The WMSN provides several advantages over traditional monitoring and surveillance systems which include [36]:

L. Ang et al., *Wireless Multimedia Sensor Networks on Reconfigurable Hardware*,
DOI 10.1007/978-3-642-38203-1_2, © Springer-Verlag Berlin Heidelberg 2013

- *Enlarging the view.* Viewpoints from multiple cameras can provide a close-up view of an event either through the images captured by a camera nearer the scene or by engaging a node with a more advanced camera such as a pan-tilt-zoom (PTZ) camera. In such a system, an event detected by a node with a lower resolution camera can signal another node with a PTZ camera to detect and track the event.
- *Enhancing the view.* The use of multiple cameras can also enhance the view of an event by providing a larger field of view (FOV) or by using cameras with different capabilities such as mixing cameras for the visible and infrared spectrum in the network. Such systems are very useful when the view is obscured or when there is little or no illumination in the scene.
- *Providing multiple viewpoints for the same event.* When a single camera is considered for a surveillance application, the coverage of the application is only limited by the FOV of a fixed camera or the field of regard (FOR) of a PTZ camera. This is limiting as parts of a scene may often be obscured especially in monitoring areas of high object density such as in public transportation hubs.

These advantages come at the cost of an increase in the data generated in the network which in turn increases the energy consumption in the network. To ensure that the typical battery-powered WMSN lifespan is not significantly affected by this, the amount of data routed through the network can be reduced with the use of in-network processing techniques such as to remove redundancy from multi-camera systems, selective transmission of the data and compressing the data. These processing tasks can be performed at the node, cluster or distributed throughout the network. The use of error detection and correction can also help reduce the likelihood of a costly retransmission. In this chapter, a broad coverage on WMSN technology will be provided as the background information for understanding the WMSN design, its architectures, challenges and design considerations. These design considerations are strongly dependent on the application. Aspects such as deployment density, cost, size, geographic location and purpose determine the components for the implementation of a specific WMSN.

2.2 WMSN Network Technology

This section will describe the layout of a typical WMSN as shown in Fig. 2.1. The network typically consists of a large number of sensor nodes deployed in a region of interest and one or more base stations. The base station or sink acts as the main network controller or coordinator. In this role, its primary function is to coordinate the functions of the nodes. It also collects information gathered by the nodes to be stored or further processed. The sink also serves as a gateway to other networks. The sinks are normally located close to the nodes to avoid high energy-consuming long-range radio communications. The energy consumption in the WMSN will determine the network lifespan. The energy consumed for communication is much higher

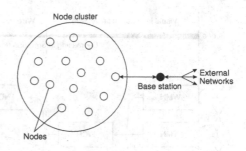

Fig. 2.1 Typical WMSN layout

than that for sensing and computation and grows exponentially with the increase of transmission distance. Therefore it is important that the amount of transmissions and transmission distance be kept to a minimum to prolong the network lifespan. This is one of the main motivations for in-network processing—the reduction of the information required for transmission for the efficient use of energy. In [60, 91], it was reported that the transmission of data can take 1,000–10,000 times more energy than processing, and this difference will increase as processor technology improves.

To reduce the transmission distance, a multi-hop short-distance communication scheme is preferred, and in most sensor networks, this is how it is implemented. In a multi-hop communication network, a sensor node transmits data towards the sink via one or more intermediate nodes. The architecture of a multi-hop network can be organised into two types: flat and hierarchical. The next sections will describe the structure of a WMSN and its various architectures and classifications.

2.2.1 Structure of a WMSN Network

Figure 2.2 shows a general structure of a WMSN consisting of four main components: wireless multimedia node (WMN), wireless cluster head (WCH), wireless network node (WNN) and base station. There is a decreasing information flow from the WMNs to the base station. The captured scene data are processed and transformed to useful event data. The focus for the upper levels of the network (WMN and WCH) is on information processing, and the focus for the lower levels of the network (WNN) is on wireless network communications. The primary theme for both upper and lower network levels is to achieve energy efficiency within the constraints of the battery-powered nodes.

2.2.1.1 Wireless Multimedia Node

The WMNs form the end points of the network. Each WMN consists of a camera or audio sensor, processing unit, communication unit and power unit. The camera

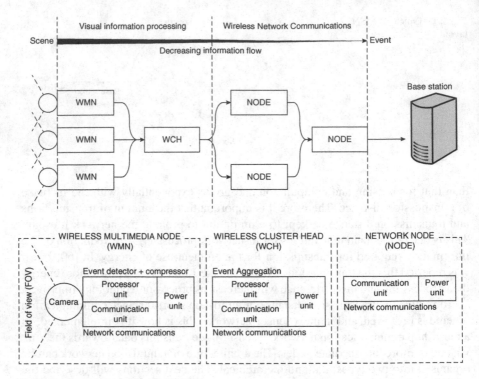

Fig. 2.2 General structure of a WMSN

sensor has a FOV of the scene. A captured scene is called an image frame. The processing unit performs the visual processing to reduce the high amount of scene data. Two approaches can be used. The first approach uses event detectors to identify useful events in the scene data. If an event is not detected, then the image frame is discarded and there is no need to transmit the frame through the network. The second approach uses event compressors to reduce the data for image frames that have to be transmitted through the network. Various image and video information processing techniques can be used, and this will be briefly described in Sect. 2.4 and discussed in detail in Chaps. 5 and 6. The communication unit transmits the compressed data to other nodes. Each WMN receives its power supply from a power unit which is mostly battery-powered.

2.2.1.2 Wireless Cluster Head

The WCHs receive data from several WMNs. Each WCH consist of a processing unit, communication unit and power unit. Each WCH receives data from several WMNs. The FOVs of WMNs may overlap, and the WCH can further reduce the data by performing event aggregation by consolidating the information from the WMNs. For example, the image frames from different WMNs may be stitched into a single frame to remove the overlapping data. This will be further discussed in Chap. 7. The communication unit and power unit perform a similar role as for the WMN.

2.2.1.3 Wireless Network Node

The WNN performs the same role as for a traditional wireless sensor network and consists of a communication unit and power unit. The communication unit relays the data from node to node until it arrives at the base station.

2.2.1.4 Base Station

The base station is the destination of all the data gathered throughout the network. This is likely to be a conventional computer capable of powerful processing and connected to a main power supply. Thus, energy efficiency and power issues are not important here.

2.2.2 WMSN Network Architecture

This section will discuss various architectures and classifications for WMSNs in terms of its composition (homogeneous or heterogeneous), its tier architecture (single-tier or multi-tier) and its mote platform architectures (lightweight-class, intermediate-class or PDA-class).

2.2.2.1 Homogeneous and Heterogeneous Architectures

A homogeneous WMSN consists of nodes that have the same capability in terms of energy, computation and storage, whereas a heterogeneous WMSN would have nodes that have differing capabilities in terms of sensing, processing and communication [5]. A heterogeneous WMSN may contain sensor nodes that are better equipped with more processing power, memory and energy storage and also better communication compared to other sensor nodes. Using better nodes such as this, the network can assign more processing and communication tasks to these sophisticated nodes (e.g. as WCHs) in order to improve its energy efficiency and thus prolong the network lifetime.

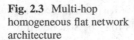
Fig. 2.3 Multi-hop homogeneous flat network architecture

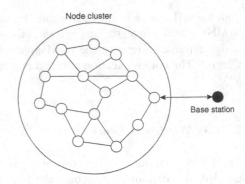

2.2.2.2 Single-Tier and Multi-Tier Architectures

In a typical WSN, the focus of the architecture design is on it being scalable. This is usually achieved through a flat and homogeneous architecture with nodes that have the same physical capabilities. An example of this is a multi-hop homogeneous flat architecture where all the nodes in the network are peers and perform the same sensing function and transmit the data to the base station or controller. This is shown in Fig. 2.3. In contrast, the WMSN is inherently heterogeneous in nature and generates different types of network traffic such as still images, video, audio or scalar data, all of which have to be supported by the architecture. The large amount of data generated by multimedia sensors may not be suited for flat architectures. Similarly, the data processing and energy requirements for communication may differ between the nodes. A node with only a microphone will generate less data and require less processing power than one which has an image sensor. These intrinsic differences of WMSNs require heterogeneous network architectures. These architectures can be classified into two categories: single-tier and multi-tier architectures. Single-tier architectures are based on a flat topology where the network can be composed of either homogeneous or heterogeneous components. Multi-tier architectures, on the other hand, exploit the higher processing and communication capabilities of high-end nodes and use a hierarchical network operation.

The hierarchical network is a type of network where the nodes are grouped into clusters as shown in Fig. 2.4. It divides a large network into smaller groups. The cluster members send their data to the cluster heads, while the cluster heads serve as relays for transmitting the data to the base station. The advantage of this is that not all raw data which are collected from the nodes need to be sent to the base station. The data can be combined together to extract only the useful information and is termed as data aggregation [106]. Another advantage of a hierarchical network is that different nodes can be used to optimise the use of energy. A low energy node can be designed to perform sensing tasks and short-range communication, while a node with higher energy can be selected as a cluster head to process the data from its cluster members and transmit the processed data to the base station.

Fig. 2.4 Hierarchical
single-tier single-hop network
architecture

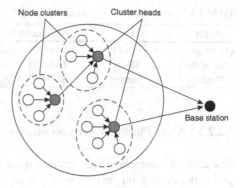

Fig. 2.5 Hierarchical
single-tier multi-hop network
architecture

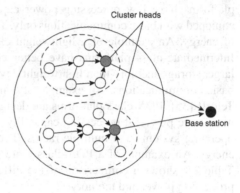

Fig. 2.6 Hierarchical
multi-tier multi-hop
architecture

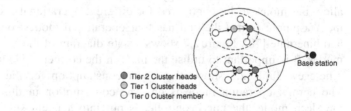

This process not only reduces the energy consumption for communication but also balances the traffic load and improves scalability when the network grows. The major problem with clustering is how to select the cluster heads and how to organise the clusters. In this context, there are many clustering strategies. Depending on the number of tiers in the clustering hierarchy, a sensor network can be organised into various architectures. Figure 2.5 shows a single-tier clustering architecture, and Fig. 2.6 shows a multi-tier clustering architecture. There have been many clustering algorithms which have been proposed to address the clustering issues.

Table 2.1 Comparison of mote platforms

Platform	Mote	Microcontroller	RAM	Flash memory
Lightweight	FireFly	8-bit ATmega128L	8 kB	128 kB
Intermediate	TelosB	16-bit TI MSP430	10 kB	1 MB
PDA-class	Stargate	32-bit PXA255 XScale	64 MB	32 MB

2.2.2.3 Mote Platform Architectures

Researchers have classified wireless mote platform architectures into three categories depending on their processing power and storage: lightweight-class platforms, intermediate-class platforms and PDA-class platforms [6]. Lightweight-class platforms have low processing power capability, small storage and are usually equipped with basic communications only. These motes consume a very low amount of energy. An example of a lightweight-class mote platform is the FireFly [86]. Intermediate-class platforms have better computational processing power and a larger storage memory than lightweight devices but are also usually equipped with basic communications. An example of a intermediate-class mote platform is the TelosB [35]. PDA-class platforms are designed to process multimedia content and have more powerful processing capability and large storage. It supports different operating systems and multiple radios. The drawback is that they consume more energy. An example of a PDA-class mote platform is the Stargate platform [34]. Table 2.1 shows a comparison of the different mote platforms in terms of their processing power and memory.

To minimise the mote energy requirements, most commercial manufacturers allow the motes to be configured for different operational modes. For example, the Waspmote from Libelium has four operational modes: on, sleep, deep sleep and hibernate [78]. Figure 2.7 shows a state diagram of the Waspmote operational modes. It is important to utilise the mote in the correct mode to maximise energy efficiency. Table 2.2 shows the power consumption for the Waspmote modes. The normal operation mode is on. The consumption in this state is 9 mA. In the sleep mode, the microcontroller is put into a latent state. It can be woken up by asynchronous interrupts and by the synchronous interrupts generated by the watchdog timeout. The duration interval of this state is from 32 ms to 8 s. The consumption in this state is 62 μA. In the deep-sleep mode, the microcontroller is put into a latent state and can be woken up by asynchronous interrupts and by the synchronous interrupts triggered by the real-time clock (RTC). The duration interval of this state is from 8 s to minutes, hours or days. The consumption in this state is 62 μA. In the hibernate mode, the microcontroller and all the Waspmote modules are completely disconnected. In this mode, it can only be woken up through the programmed alarm in the RTC. The duration interval of this state is from 8 s to minutes, hours or days. The consumption in this state is only 0.7 μA.

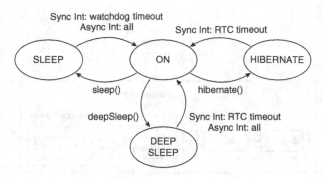

Fig. 2.7 State diagram for Waspmote operational modes

Table 2.2 Operational modes for Waspmote

	Consumption	μProcessor	Cycle	Accepted interruptions
ON	9 mA	ON	–	Synchronous and asynchronous
Sleep	62 μA	ON	32 ms–8 s	Synchronous (watchdog timeout) and asynchronous
Deep sleep	62 μA	ON	8 s–min/h/d	Synchronous (real-time clock) and asynchronous
Hibernate	0.7 μA	OFF	8 s–min/h/d	Synchronous (real-time clock)

2.3 WMSN Node Technology

A notable difference between the WMSN and the WSN on which it is based on is its multimedia handling capabilities. Typical WSNs are not designed for multimedia tasks, but embedding audio and video capture devices on the WSN enhances its capabilities and broadens its potential applications. The primary challenge for the WMSN hardware architecture and its components is the need to support the higher bandwidth and processing requirements needed for audio sensors such as microphones and low-resolution and medium-resolution image sensors. Low-resolution image sensors can be built specifically for WMSNs and are part of an embedded system which may consist of low-end transceivers as well as specialised processor and memory units that are required for low-power image processing and storage.

2.3.1 Structure of a WMSN Node

The structure of a WSN node is designed to be physically small, low cost and with low power consumption. A typical node consists of four main components:

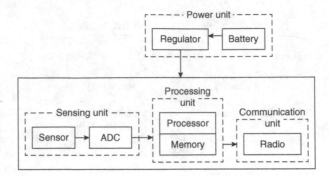

Fig. 2.8 Components of a typical WSN node

sensing unit, processing unit, communication unit and power unit. Figure 2.8 shows
the components for a typical WSN node.

- The sensing unit consists of one or more sensors and analog-to-digital converters
  (ADCs). The sensors detect and measure the physical occurrences which are
  analog in nature. This analog signal is converted to a digital signal using an ADC.
  This digital signal is then input to the processing unit.
- The processing unit usually consists of a microcontroller or microprocessor with
  memory as part of the same die or integrated circuit package. The processing unit
  could also be or contain a digital signal processor (DSP) or application-specific
  processor. The processing unit provides intelligent control and processing of
  information to the sensor node.
- The communication unit consists of a short-range transceiver, typically based
  on the IEEE 802.14.3 or ZigBee™ standard. There are also many other wireless
  communication protocols which could be used.
- The power unit regulates power to and from an energy storage unit. This energy
  storage unit is typically a compact and portable storage such as a battery.
   In most configurations, the ADC, processor, memory and radio are integrated
  into a single system on a chip (SoC). The details of each of these units will be
  discussed in the following sections.

2.3.2   Sensing Unit

The sensing unit comprises of the sensors and supporting circuits to capture
data from the area or environment being monitored. Examples of scalar sensors
are those which monitor physical environmental conditions such as temperature,
pressure, humidity, light or sound pressure levels. These sensors are typically

Table 2.3 SHT1x, 2x and 7x series of humidity sensors from Sensirion

Sensor model		Max. tolerance			Sensor output
		Packaging	RH	T	
SHT10		SMD	±4.5%RH	±0.5°C	Digital Sbus
SHT11		SMD	±3%RH	±0.4°C	Digital Sbus
SHT15		SMD	±2%RH	±0.3°C	Digital Sbus
SHT21		DFN	±3%RH	±0.4°C	I²C, PWM, SDM
SHT25		DFN	±2%RH	±0.3°C	I²C
SHT71		Pins	±3%RH	±0.4°C	Digital Sbus
SHT75		Pins	±1.8%RH	±0.3°C	Digital Sbus

low-powered passive devices. Examples of multimedia sensors are electromagnetic wave detectors such as visible spectrum, infrared and ultraviolet image detectors and acoustic detectors. The sensed data collected is analog in nature and is usually amplified before digitisation by an analog-to-digital converter (ADC). This can be performed by either the sensor package or the processor unit. The earlier is preferred as low-level analog signals are kept within the sensor package minimising possible external interference and loss, and the amplifier, filter and ADC can be better matched to the capabilities and characteristics of each sensor.

2.3.2.1 Scalar Sensor

A common example of a scalar sensor application is to sense the temperature and humidity of an environment. These can be achieved through the use of a combined sensor such as the humidity sensors from Sensirion [112]. Table 2.3 shows the Sensirion series of humidity sensors. These integrated digital sensors combine a temperature and humidity sensor with an on-chip calibration memory, ADC and digital interface in a complementary metal oxide semiconductor (CMOS) design. This helps ensure the sensors are small (the SHT25 measures $3 \times 3 \times 1.1$ mm) and easy to interface via the digital interface.

2.3.2.2 Image Sensor

The image sensors found on WMSNs are typically CMOS sensors. Figure 2.9 shows the structure for a (Bayer pattern) CMOS sensor. These CMOS sensors are produced in the same manner as computer-integrated circuits. The high production volume of such devices and the high yield through continuous improvements in CMOS manufacturing processes ensures that these sensors remain low cost, small in size and ideal for use in WMSNs. By integrating the photosites, amplifiers and image processing circuits on the same chip, CMOS sensors consume much lower

Fig. 2.9 A typical CMOS
sensor arrangement (Bayer
pattern filter)

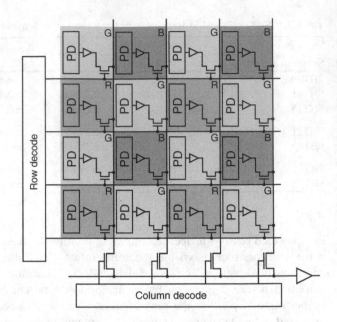

energy than charge-coupled device (CCD) image sensors. These sensors also benefit
from the continuous improvements in CMOS manufacturing scale reduction by
allowing the supporting circuits to be smaller. This allows for more space on the
chip for photosites which leads to better low-light performance or further sensor
miniaturisation. Recent backlighted (or back-illuminated) CMOS (BSI) sensors
offer much better low-light performance for a given size than their non-backlighted
counterparts. BSI technology involves turning the image sensor upside down and
applying colour filters and micro-lenses to the backside of the pixel so that light is
collected through the backside of the sensor. It reverses the arrangement of layers
so that metal and dielectric layers reside below the sensor array, providing the
most direct path for light to travel into the pixel, which optimises the fill factor
to deliver best-in-class low-light sensitivity, image quality and colour reproduction.
This approach differs from conventional front side illumination (FSI) architectures,
where light travels to the photosensitive area through the front side of the pixel.
This requires the light to first pass through transistors, dielectric layers and metal
circuitry, which can block or deflect it into neighbouring pixels, causing a reduced
fill factor and additional problems such as cross talk between pixels. An example of
a BSI sensor is the Omnivision OV16820 [98] which is a 16-megapixel chip capable
of capturing 30 fps, with a manufacturer claimed power consumption of only 10 µA
in standby and 310 mA when active. This sensor outputs the captured image as a
10- to 12-bit RAW RGB image or a 8- to 10bit differential pulse-code modulation
(DPCM) compressed image. Figure 2.10 shows the difference between a FSI and
BSI CMOS sensor architecture.

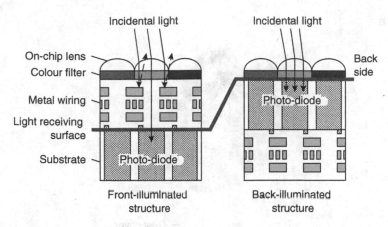

Fig. 2.10 Front-illuminated vs. back-illuminated CMOS sensor

2.3.2.3   Audio Sensor

Audio sensors on WMSNs have to be low powered and small in size. These sensors vary in size, cost and application. Other environmental factors such as vibration, humidity and temperature also need to be taken into consideration when selecting an audio sensor. The typical audio sensor consists of a microphone, amplifier (can be several stages), filters and an ADC. Generally, there are two types of audio sensors which are suited for WMSNs. These are either based on the electret condenser microphone (ECM) or the microelectrical-mechanical system (MEMS) microphone. In some monitoring applications such as structure or traffic monitoring, an audio sensor could be very useful as a sentry. The sensor node can be in a deep-sleep state to conserve energy and wakes up only when an audio threshold is exceeded such as from a collision or from structural stresses and strain. Figure 2.11 shows the structure of a the ECM. .

The ECM is a type of condenser microphone with a permanently charged electret material fixed to the capsule back-plate. It eliminates the need for a polarising power supply as required by condenser microphones. ECM microphones provide a good dynamic range and are found in many applications from high-quality recording to built-in microphones in telephones. It is widely available, low cost and small enough to be embedded into many applications. The MEMS microphone is a solid-state microphone that typically comes integrated with a matching ADC, filter, power management, hardware control and communication port. A major advantage of the MEMS microphone over the ECM microphone is that it is able to interface to the node processor over a digital serial link. Other advantages of the MEMS are as follows:

- *Higher performance density.* For the same volume, MEMS microphones can achieve better noise performance than equivalent ECMs.

a

b

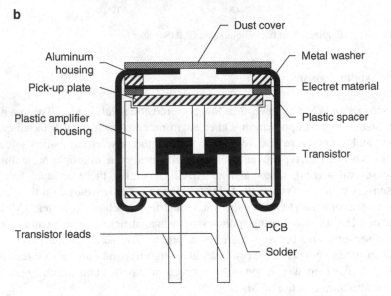

Fig. 2.11 The ECM microphone (foil type or diaphragm type) (**a**) Photo of ECM. (**b**) Cross-section of ECM

- *Less variation in sensitivity over temperature.* The sensitivity of an ECM may drift as much as $\pm 4\,$dB over its operating temperature range, while an MEMS microphones sensitivity may only drift by 0.5 dB over the same range.
- *Lower vibration sensitivity.* The mass of a MEMS microphone diaphragm is lower than that of an ECM, leading to less response to vibrations in a system. This is important in applications where vibrations are a common occurrence.
- *More uniform part-to-part frequency response than ECMs.* A random selection of MEMS microphones of the same type will have near identical responses. This is important in microphone arrays such as those made possible using WMSNs. Otherwise, different sensor nodes may capture different information from the same event.

Fig. 2.12 Processor unit

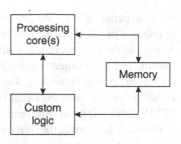

- *Can be reflow soldered.* This saves cost by allowing assembly on the same reflow soldering process as most other integrated circuits on a printed circuit board (PCB) and helps to reduce the manufacturing costs.
- *Semiconductor manufacturing process.* Like the CMOS image sensors, the solid-state MEMS microphone also benefits from the advancements in semiconductor manufacturing processes.

2.3.3 Processor Unit

The WSN processor unit consists of very low-powered and deep-sleep-capable processing cores. Figure 2.12 shows an architecture of a processor unit. The processor unit consists of at least one processing core and local memory. In some configurations, it may also contain custom logic which is more efficient in terms of the processing/power consumption ratio over using a general-purpose processor (GPP).

2.3.3.1 Processing Core

Processor architectures can be classified into three categories: GPPs, DSPs and application-specific instruction set processors (ASIPs). These processor architectures will be discussed in the following sections.

General-Purpose Processors

A typical WSN will contain at least a basic processor for coordinating the functions on the WSN. This processor can be termed as a GPP and needs to be sufficient for making simple computations, serialisation and packetisation of data for transmission. It is also responsible for decoding and interpreting incoming messages. To conserve energy, the processor should have the following features: a deep-sleep state with low-current wake-up, a scalable clock and a simple instruction set. However,

the emergence of WMSNs has changed the role of the processor from its basic role
to one that has to process more complex multimedia data such as to pre-process and
compress the data from the multimedia sensors before transmission. To meet the
multimedia requirements, a different kind of low-power processing core with more
processing power is needed. Another consideration in choosing the processing core
is that it needs to have enough I/O interface which is usually serial in nature such as
SPI (Serial Peripheral Interface Bus), I^2C (Inter-Integrated Circuit) or 1-wire. This
is required for connecting various components of the system such as the external
memory, ADCs and radio. Examples of GPP processors are the Atmel AVR and
those based on the ARM [12] and MIPS [95] architectures.

Digital Signal Processors

Higher-end WMSN sensor nodes may also contain a digital signal processor (DSP).
DSPs are specialised processor architectures to exploit data and instruction-level
parallelism in signal and image processing algorithms. Although DSPs in general
can use a fixed-point or floating-point arithmetic format, DSPs for WMSNs would
use a fixed-point format to be more energy and area efficient. DSPs are often
optimised to perform operations such as digital filtering or fast Fourier transform
(FFT) operations. A typical DSP would be able to achieve a throughput of one
clock cycle per filter tap. It is able to achieve this using a Harvard computer
architecture with multiple registers and specialised hardware like circular buffering and multiply-and-accumulate (MAC) units where multiple operations can be
performed in a single clock cycle. A Harvard architecture would contain separate
memories and buses for instructions and data storage termed Instruction Memory
and Data Memory, respectively. This allows a new instruction to be fetched from the
Instruction Memory in the same cycle as data is accessed from the Data Memory.
The circular buffer is a data structure in memory which is used to store incoming
data. When the buffer is filled, the incoming new data is written at the beginning
of the buffer and overwrites the old data. DSPs also use pipelining architectures
to improve the execution throughput where the operations of the instruction cycle
are split into smaller sequential tasks that are executed in parallel by different units
within the processor.

Application-Specific Instruction Set Processors

In contrast to GPPs and DSPs which target a broad range of applications, ASIPs are
optimised to target a single application. ASIPs are often implemented by customising the instruction set architecture (ISA) of a GPP processor. It takes advantage
of user-defined instructions and a user-defined datapath which is optimised for a
particular application. The resulting optimisation achieves a higher computational
performance and energy efficiency than a GPP processor. Examples of ASIP

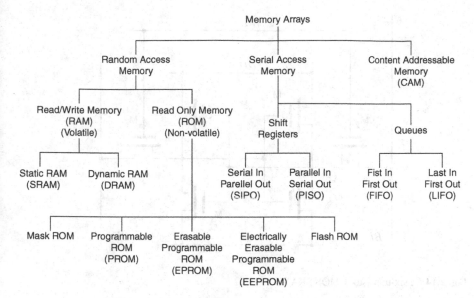

Fig. 2.13 Different memory types

processor customisations for WMSNs are for applications involving encryption, error correction and compression. The work in [127] describes an ASIP processor for encryption. In this work, the authors describe four ASIP designs (8-bit, 32-bit, 64-bit and 128-bit ASIP) for optimising the Advanced Encryption Standard (AES) for implementation in a WSN. The authors performed a comparison among the four cryptographic designs in terms of performance, power consumption, energy dissipation and area occupation and concluded that the smallest ASIP (8-bit design) is the most power efficient.

2.3.3.2   Local Memory

The local memory found in a WMSN is either used to store program codes or sensed data. The program codes are normally stored on non-volatile memory built into the processor core. Sensed data such as those from an image sensor is normally stored in an external memory. There are typically two types of memory used: random access volatile and non-volatile. Figure 2.13 shows the different memory types. Random access volatile memory describes memory that requires power to maintain the stored information, whereas non-volatile memory does not require a maintained power supply. Dynamic random access memory (DRAM) is a random access memory that stores each bit of data in a separate capacitor. This is commonly found in applications where large amounts of memory are needed and power is not a major concern. For most embedded systems such as that used in WMSNs, the use of DRAM seems impractical as it requires a refreshing circuitry because

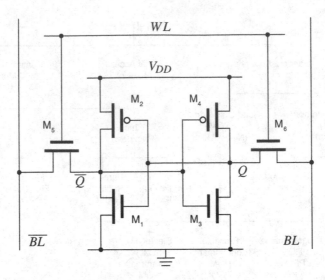

Fig. 2.14 A six-transistor CMOS SRAM

the capacitors it uses lose charge over time. The need for a refresh cycle increases the integration complexity and decreases the overall performance and can also lead to higher power consumption. However, it is relatively much cheaper compared to static random access memory (SRAM) and is available in large sizes with modules in the gigabyte range.

SRAM uses a bistable latching circuitry to store each bit and does not require a refresh circuit. Figure 2.14 shows a CMOS SRAM cell which uses six transistors. This is configured as four transistors to store a bit and another two transistors for control access. Furthermore, this design can be extended to implement more than one read/write port (termed as a multiport memory), allowing the memory to be read and written within the same clock cycle. SRAMs are more expensive to manufacture and are typically found in smaller sizes from a few kB to 512 Mbit. Also available are cheaper pseudo SRAM (e.g. Winbond W968D6DKA [136]) which works like SRAM but is built on a DRAM core using capacitors with an integrated on-chip refresh circuit.

Another type of memory commonly found on many embedded systems is the non-volatile memory. These are typically NOR- or NAND-based flash based on their gate types. For many systems NAND flash memory provides a low-cost and large capacity upgradable memory and is commonly available as Secure Digital Cards (SD) and Multimedia Cards (MMC) which uses SPI to communicate. These can be used as a backup memory or used to store large amounts of data that do not require fast access as these are not random access. In the NOR-based flash, the individual memory cells are connected in parallel, which enables the device to achieve random access. This configuration enables the short read times required for the random access of microprocessor instructions. NOR flash memory is ideally

Table 2.4 Comparison of NAND and NOR flash memory operating specifications

	SLC NAND flash (x8)	MLC NAND flash (x8)	MLC NOR flash (x16)
Density	512 Mbit–4 Gbit	1 Gbit–16 Gbit	16 Mbit–1 Gbit
Read speed (MB/s)	24	18.6	103
Write speed (MB/s)	8	2.4	0.47
Erase time (ms)	2.0	2.0	900
Interface	I/O indirect access	I/O indirect access	Random access
Application	Program/data mass storage	Program/data mass storage	eXecuteInPlace

used for lower-density, high-speed read applications, which are mostly read only and often referred to as code-storage applications. The characteristics of NAND flash memory are its high density, medium read speed, high write speed, high erase speed and an indirect or I/O-like access. The characteristics of NOR flash are lower density, high read speed, slow write speed, slow erase speed and a random access interface. Due to the lower density nature of NOR memory, it is also more expensive and require more power. Table 2.4 shows a comparison between the NAND and NOR flash memory operating specifications.

2.3.3.3 Custom Logic

The custom logic located in the processing unit handles certain specific tasks that have been earlier determined for the application. They help to augment the processing core. In WMSNs such tasks would be basic image and audio processing tasks. Examples for image processing tasks include edge detection, colour conversions, mathematical transforms and image compression, and examples for audio processing tasks include peak detection and audio compression. Custom logic can be implemented via an ASIC or a low-power FPGA such as the Actel IGLOO.

2.3.4 Communication Unit

The communication unit is responsible for all incoming and outgoing transmissions from the sensor node. Figure 2.15 shows several typical communication methods used in WSNs; however, it is most common to use radio frequency (RF) waves. Three common standards can be used for RF communication: IEEE 802.11 wireless LAN (normally termed Wi-Fi), Bluetooth and IEEE 802.15.4 (normally termed ZigBee™). Table 2.5 shows a comparison of the RF communication standards. Although the use of RF-based communications is common for WMSNs, it is

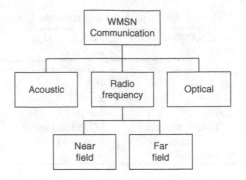

Fig. 2.15 Communication types

Table 2.5 Comparison of RF communication standards

Market name	Wi-Fi™	Bluetooth™	ZigBee™
Underlying standard	802.11b	802.15.1	802.15.4
Application focus	Web, email, video	Cable replacement	Monitoring & control
Battery life (days)	0.5–5	1–7	100–1,000+
Network size	32	16	100s to 1,000s
Bandwidth (kbits/s)	11,000+	720	20–250
Range (m)	1–30+	1–10+	1–1,000+
Network architecture	Star	Star	Mesh
Optimised for	Speed	Low cost, convenience	Reliability, low power, low cost, scalability

not limited to RF-based communications only and can be implemented using optical- or acoustical-based techniques which may be more applicable in some environments. For example, underwater WMSNs can use acoustical-based communications. Another benefit of non-RF-based communications is that the deployment of the WMSN is eased as it does not require any regulatory approval or licences unlike RF-based communications.

Recently, a carrierless RF transmission system called ultra-wide-band (UWB) technology has been proposed to meet the requirements for use in WSNs [145]. UWB is a wireless radio technology designed to transmit data at very high bandwidths (~480 Mbps) within short ranges (~10 m) using little power. The UWB has been added to the IEEE 802.15.4a standard as an additional physical layer. Other benefits of the carrierless UWB is that it offers a simplified radio compared to narrowband communication and the large bandwidth offered by the UWB helps in node positioning by allowing centimetre accuracy in ranging [60]. The large bandwidth also decreases the power spectral density and reduces the interference to other systems. The next four sections will describe three non-RF-based communication methods, optical, acoustic and magnetic, followed by a description on the RF-based

ZigBee standard. Typically, the low data rate (max of 250 kbits) of ZigBee is not feasible for use in WMSNs, but advancement of in-node processing techniques has enabled low data rates suitable for use in ZigBee networks.

2.3.4.1 Optical Communications

Optical communication takes place by encoding data into pulses of light (visible or otherwise) which is typically from the lower end of the electromagnetic spectrum wavelength of 600–1.000 nm This is done by modulating the light from a transmitter such as a light-emitting diode (LED) or a laser diode (LD) which can be captured by a photodiode at the receiver end and decoded. The result is an intensity modulation with direct detection (IM/DD) system. This system can be implemented over a solid light guide (e.g. optical fibre cable) or through free space (e.g. air or vacuum). A common form of optical communication is to use infrared communication. Infrared communication has been used mainly for short-range communications in portable devices such as cell phones, laptops and PDAs. This has been replaced by alternative RF-based methods like Bluetooth which does not require line-of-sight (LoS). The LoS point-to-point communication is the most commonly used IR communication type. Another type is diffuse IR which does not require LoS but works by reflecting the light from surrounding objects. This would normally require a wide angle transmitter and receiver. An example of an implementation of optical communication is the Smart Dust mote [70, 134].

2.3.4.2 Acoustic Communications

Acoustic communications are used only when both RF and optical transmission options are not feasible such as in an underwater environment. A low frequency (30 Hz–10 kHz) acoustic signal can propagate in water over very long distances. However, there are several issues to be overcome such as attenuation and noise, multipath propagation and the Doppler effect. Sound propagates underwater at a very low speed of 1.500 ms, and the propagation occurs over multiple paths. The delay spreading over tens or even hundreds of milliseconds results in a frequency selective signal distortion while motion creates an extreme Doppler effect [117]. All of these depend on the properties of the water such as temperature, depth and composition.

2.3.4.3 Near-Field Radio Frequency Communications

In a far-field RF communication system, most of the energy is contained in the electric field. Near-field RF communication makes use of the strong magnetic fields in the near field of the signal but weak in the far field. In essence, a

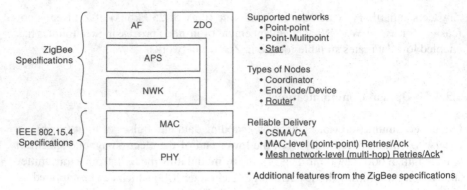

Fig. 2.16 ZigBee layers and specifications

near-field magnetic induction (NFMI) communication system is a short-range wireless physical layer that communicates by coupling a tight, low-power and non-propagating magnetic field between devices. The system uses a transmitter coil in one device to modulate a magnetic field which is measured by means of a receiver coil in another device. NFMI systems are designed to contain transmission energy within the localised magnetic field. This magnetic field energy resonates around the communication system but does not radiate into free space. The standard modulation schemes used in typical RF communications (amplitude modulation, phase modulation and frequency modulation) can be used in NFMI systems.

2.3.4.4   Far-Field Radio Frequency Communications

The objective of the IEEE 802.15.4 standard was to provide a wireless communications standard for ultra low complexity, ultra low cost, ultra low power consumption and low data rate (maximum of 250 kbits) wireless connectivity among inexpensive devices such as alarms, sensors and automation devices. The standard defines the physical and medium access control layers that specify a low rate personal area network (LR-PAN). Although, there are several protocols which uses the 802.15.4 as its underlying layers such as the Wireless HART [57], ISA-SP100 [122] and IETF IPv6-LoWPAN [96], the most widely known is the ZigBee standard [146].

ZigBee Network Layers

Figure 2.16 shows the ZigBee layers and specifications. ZigBee operates in the industrial, scientific and medical (ISM) radio bands. The original IEEE 802.15.4 (2003) standard specifies the physical (PHY) and medium access control (MAC) layers at the 868 MHz, 915 MHz and 2.4 GHz ISM bands. Both contention-based and contention-free channel access methods are supported. Data transmission rates vary from 20 to 250 kbits. ZigBee builds upon the PHY and MAC defined in IEEE standard 802.15.4 for low-rate WPANs. The specification goes on to complete the standard by adding four main components: network layer, application layer, ZigBee device objects (ZDOs) and manufacturer-defined application objects which allow for customisation and helps integration. The ZDOs are responsible for a number of tasks which include keeping of device roles, management of requests to join a network, device discovery and security. The air interface is direct-sequence spread spectrum (DSSS) using BPSK for 868 and 915 MHz and O-QPSK for 2.4 GHz. The access method in IEEE 802.15.4-enabled networks is carrier sense multiple access with collision avoidance (CSMA-CA). The IEEE 802.15.4 PHY includes receiver energy detection (ED), link quality indication (LQI) and clear channel assessment (CCA). Table 2.6 shows a summary of the ZigBee PHY layer.

ZigBee Network Nodes

The ZigBee specification defines three node types: coordinator, router and end device. Table 2.7 shows the tasks performed by the different network node types. These nodes can be configured into three different topologies: tree, star or mesh as shown in Fig. 2.17. Every ZigBee network is controlled, initiated and maintained by a coordinator, and there is only one coordinator in a network. The coordinator node is tasked with the network creation, the control of its parameters and basic maintenance. Within star networks, the coordinator must be the central node. In a star network, the end devices communicate directly with the coordinator. Using this topology, up to 65,536 end devices can be supported using a 16-bit address. The star network has a simple layout and has low latency. In a tree and mesh network, data is passed through routers before arriving at the coordinator. Because ZigBee nodes can go from sleep to active mode in 30 ms or less, the latency can be low and devices can be responsive, particularly compared to Bluetooth wake-up delays, which are typically around 3 s. Because ZigBee nodes can sleep most of the time, the average power consumption can be low, resulting in long battery life.

2.3.5 Power Unit

The power unit of a WMSN node is responsible for regulating power from the node power source. The power unit must be able to supply the peak current demands of the processor when transitioning from a deep-sleep state to a fully operational state and during radio communications. The node power source is typically a cell (either primary or secondary) and can be augmented with an energy-scavenging system such as a photovoltaic (PV) panel which extracts solar energy or fluid turbines which extract energy from fluid flow such as a wind or wave turbine. A cell generates

Table 2.6 ZigBee PHY layer [IEEE 802.15.4 specifications, 2003–2012]

Band (MHz)	Region	Channel		Center frequency (MHz)	Modulation type		Bit rate (kb/s)	Notes
		Page	Num					
780 (c2009)	China (802.15.4c-2007) 8 channels	5	0–3	780 (2 MHz channel spacing)	DSSS	M-ary PSK	250	• 4-bits per symbol • 16-chip PN • Raised cosine pulse shaping
			4–7			O-QPSK	250	
868	Europe	0	0	868.3 (2 MHz channel spacing)	DSSS	BPSK	20 (2003)	• 15-chip PN • Raised cosine pulse shaping
		2	0			O-QPSK	100 (2006)	
(2006)		1	0		20-bit PSSS	ASK	250	
915	USA and Australia 30 channels	0	1–10	906–954 (2 MHz channel spacing)	DSSS	BPSK	40 (2003)	
		2	1–10			O-QPSK	250 (2006)	
		1	1–10		5-bit PSSS	ASK	250 (2006)	

950 (d2009)	6	Japan (802.15.4d-2009) 22 channels	0–7 (1 mW)	951.2–955.4 (0.6 MHz channel spacing)	DSSS	BPSK	20	
			8–9 (10 mW)	954.4–954.6 (0.2 MHz channel spacing)				
			10–21	951.1–955.5 (0.4 MHz channel spacing)		GFSK	100	
2,400	0	Worldwide 16 channels	11–26	2,405–2,480 (5 MHz channel spacing)	DSSS	BPSK	250 (2003)	• 32-chip PN • Half sine pulse shaping

PSK phase-shift keying, *DSSS* direct-sequence spread spectrum, *BPSK* binary phase-shift keying, *O-QPSK* offset quadrature phase-shift keying, *ASK* amplitude-shift keying, *PSSS* parallel sequence spread spectrum, *GFSK* Gaussian frequency-shift keying

Table 2.7 ZigBee and 802.15.4 network node types

Node type		
802.15.4	ZigBee™	Tasks
Full function device (FFD)	Coordinator	The coordinator is the "master" device; it governs all the network nodes • One per PAN • Establishes and organises the PAN • Mains-powered (typically)
	Router	Routers route the information which sent by the end devices • Optional in a network • Several can be in a PAN • Mains-powered (typically) • Can serve as motes
Reduced function device (RFD)	End device, end nodes, motes	These are the sensor nodes, the ones which take the information from the environment • Several can be in a PAN • Low power-, mostly in a deep-sleep state • Battery-powered (typically)

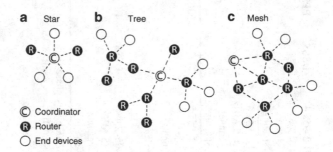

Fig. 2.17 ZigBee network topologies

electrical energy through a electromechanical conversion from stored chemical energy. A cell is typically known by its chemistry type such as the more commonly used primary cells such as the zinc-carbon or alkaline (zinc-manganese dioxide). For secondary cells, these are lead acid, nickel cadmium (NiCd), nickel metal hydride (NiMH), lithium ion (Li-ion) and lithium ion polymer (Li-ion polymer). In systems with both a secondary cell and an energy-scavenging system, the power unit must also be able to regulate the charging of the cell. There are several considerations when selecting a cell type for a node. Most important is the environment or conditions it will operate such as the operating temperature, vibration, transient and steady-state current drain of the node. For instance, lithium ion cells have a diminished capacity at subzero temperatures.

2.3.6   Frame Memory

The frame memory is a memory space used to temporarily store the image (or a sequence of images). It is similar to that of the local memory except that it serves only one purpose which is as an image buffer. This is needed as the image capture rate can be higher than the processing ability of the node.

2.3.7   Localisation Unit

In many applications, the location or position of the nodes whether in relation to each other or their absolute positions are important. The gathered information needs to be associated with the location of the sensor nodes to provide an accurate view of the observed sensor field. This is especially true if the events being monitored is time varying or if the nodes are mobile. Positioning systems can be divided into three main categories: time-of-arrival (TOA), angle-of-arrival (AOA) and signal-strength (SS)-based systems. The TOA technique estimates the distance between nodes by measuring the travel time of the received signal. The AOA technique measures the angles between a given node and a number of reference nodes to estimate the location. The SS approach estimates the distance between nodes by measuring the energy of the received signal. One of the most commonly used positioning system GPS is time-based. Trilateration is used to determine the position by the measurement of distances. These distances are obtained by measuring the travel times of signals between nodes. If two nodes have a common clock, the node receiving the signal can determine the TOA of the incoming signal that is time stamped by the reference node. The main challenge in such systems is ensuring that the clock is synchronised throughout the system as it affects the TOA estimation accuracy. Many outdoor deployments of WMSNs which require positioning are fitted with a GPS localisation unit for positioning as they are relatively low cost, available off-the-shelf and easy to implement. Most units return the position in the NMEA 0183 (National Marine Electronics Association) format and can be as accurate as pm2 m.

In indoor locations, GPS signals are often too weak for providing positioning information. However, this has been somewhat overcome by the use of massively parallel correlation processing that makes it possible to fix position with GPS signals as weak as -150 dBm [54]. One possibility is to use radio signals from the WMSN units, but localisation of radio signals indoors is difficult because of the presence of shadowing and of multipath reflections from walls and objects. By using UWB radios, the wide bandwidth of its signals implies a fine time resolution that gives them a potential for high-resolution positioning applications using the time-difference-of-arrival (TDOA) method provided that the multipaths are dealt with. Due to the short-burst and pulse-like nature of UWB signals, multipaths are easily detected and overcome by the correlator in the multipath combining receiver.

Fig. 2.18 Position
calculation using
angle-of-arrival

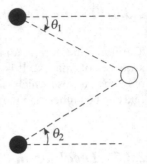

An AOA-based positioning technique involves measuring angles of the target node
seen by reference nodes (triangulation) which is done by means of antenna arrays.
To determine the location of a node in a two-dimensional (2D) space, it is sufficient
to measure the angles of the straight lines that connect the node with the two
reference nodes as shown in Fig. 2.18.

The AOA approach is not suited for implementation in WMSNs because it
requires the use of expensive antenna arrays and WMSN nodes typically use
a low-cost omnidirectional antenna. Furthermore, the number of paths may be
very large, especially in indoor environments. Therefore, accurate angle estimation
becomes very challenging due to scattering from objects in the environment. The SS
positioning estimation relies on a path loss model. The distance between two nodes
can be calculated by measuring the energy of the received signal at one node. This
trilateration technique is similar to the TOA technique but uses the SS or received
signal strength indicator (RSSI) for distance estimation rather than travel time. To
determine the distance from SS measurements, the characteristics of the channel
must be known. SS-based positioning algorithms are very sensitive to the accurate
estimation of those parameters.

2.4 Information Processing in WMSN

To realise the full potential of the WMSNs, the nodes can collaborate with each
other to achieve the goals of the application. This section will discuss various issues
for collaborative information processing in WMSNs such as multi-camera network
models, collaborative object detection, tracking and recognition and information
reduction techniques for WMSNs.

2.4.1 Multi-Camera Network Models

Researchers have classified multi-camera networks into three models: centralised
processing, distributed processing and clustered processing [4]. Figure 2.19 shows

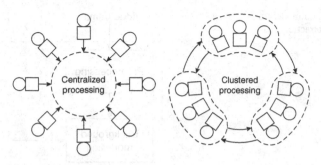

Fig. 2.19 Multi-camera network models

the centralised and clustered processing models. The centralised processing model resembles a traditional WSN architecture where the data sensed by the nodes are sent to the base station for specific application processing. The processing performed in the nodes are only for energy efficiency requirements, for example, compression. The clustered processing model allows the nodes to perform the visual processing at the node level. Each cluster may also collaborate and share data with its neighbouring clusters. The distributed processing model is a clustered processing model where each cluster only consists of a single node.

2.4.2 Collaborative Object Detection, Tracking and Recognition

Typical visual processing tasks to be performed collaboratively are to detect, track and recognise objects in the scene. Object detection refers to a visual processing task to locate a certain class of objects in the image frame (e.g. humans and vehicles). The goal of object tracking is to associate target objects in different video frames. Object recognition refers to a visual processing task where the general class of the object is known and the objective is to recognise an objects exact identity. There are two approaches which can be used for object detection: detection of objects using static features and detection of moving objects. The first approach, where the object is detected using features from a single frame, involves maintaining a model of the objects of interests. The detection process is then decomposed into two steps of first finding the features in the image and then validating whether the features sufficiently explains the presence of the object. This approach has the disadvantages of being computationally intensive and also requiring training of the detection system beforehand. The second approach involves detection of moving objects. A common approach is to use a technique termed as background subtraction. Figure 2.20 shows a general model for background subtraction [26]. There are four stages in the model: pre-processing, background modelling, foreground detection and data validation. In the initial stage, smoothing operations for temporal or spatial

Fig. 2.20 General model for background subtraction

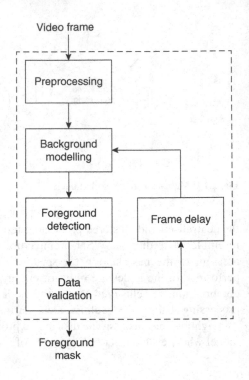

pre-processing can be performed to reduce camera and environmental noise such as rain and snow. Frame size reduction can also be performed to reduce the data processing rate. Background modelling is then performed to use the new video frame to calculate and update the background model. Various models can be used. These range from simple frame differencing to more sophisticated models using mixture of Gaussians.

The aim of object tracking is to generate the trajectory of an image over time by locating its position in every frame of the video. There are two approaches which can be used for object tracking: deterministic approaches or stochastic approaches. Deterministic approaches pose tracking as an optimisation problem, whereas stochastic approaches estimate the posterior distribution of the object location using Kalman filters or particle filters. The authors in [88] present an object tracking algorithm suitable for a network of wireless cameras. The algorithm uses a clustering protocol to establish connections among cameras that detect the target and to enable the propagation of the state of the target as it moves. Data aggregation is carried out using a decentralised Kalman filter. They carry out a series of experiments and show that the algorithm is capable of tracking a target with accuracy comparable to that achieved by a centralised approach wherein every measurement is transmitted to the base station. There are two approaches which can be used for object recognition: feature-based approaches and global approaches. Feature-based approaches detect points of interest in each image of an object and describe the object using descriptors of these local feature points. Global approaches describe the object using their global properties such as shape, silhouette, texture or colour.

2.4.3 Information Reduction in WMSN

In general, two different approaches can be employed to reduce the image data. The approaches can be divided into single-view approaches and multi-view approaches. Single-view approaches attempt to reduce the image data from each individual WMN. Two techniques can be used here: event detection and event compression. The first technique uses event detection to reduce data by only transmitting frames in the network when significant events are detected. Image frames which are not significant are discarded by the WMN. The second technique is to perform event compression by performing compression on the image frames which have to be transmitted. Event compression approaches rely on image and video compression algorithms to remove redundancy from the data. These algorithms range from current standards like JPEG, MPEG-x and H.26x to newer techniques using distributed video coding and compressive sensing. While single-view approaches attempt to reduce the scene data from each individual WMN, multi-view approaches perform the data reduction by aggregating the data from different WCHs. This is possible because of the overlapping FOVs of the different WMNs.

2.4.3.1 Event Detection

Event detection approaches reduce scene data by only transmitting image frames when significant events are detected for the application. For example, a surveillance application could use a face detector to decide which image frames to send to the base station. However, the face detector would need to have low computational complexity to meet the energy requirements in the WMN. There is a trade-off between energy required for processing and energy required for transmission. On the one hand, using an event detector in the WMN requires more computational power. On the other hand, this could result in a saving of transmission power when frames are discarded. The other advantage of an event detector is that it could also serve as an early stage for visual pre-processing. To perform the facial recognition process, the central computer would need to perform at least two stages. The first stage is to locate the face location in the image, and the second stage would then perform the recognition task by comparing the facial features with a stored database. To reduce the large amount of image data for processing by the central computer, the event detector performs the face detection task, and the location of the face is then communicated to the central computer to perform the facial recognition task. An example of a WMSN employing an event detector can be found in the paper by [37]. The authors propose an event detector using simple image processing at the camera nodes based on difference frames and the chi-squared detector. Their algorithm performs well on indoor surveillance sequences and some, but not all, outdoor sequences. The research challenge is to find suitable detectors which are reliable and can be efficiently implemented within the hardware constraints of the WMSN. Further discussions of using event detection in WMSNs are given in Chap. 5.

2.4.3.2 Event Compression

There is a wide range of image compression techniques. The research challenge is to identify algorithms which are suitable for implementation in WMSNs. As a main characteristic, the algorithms should have low computational complexity and memory requirements while maintaining high coding performance to achieve energy efficiency. In general, image coding can be categorised under first generation and second generation image coding approaches and is carried out in two steps. The first step converts the image data into a sequence of messages. In the second step, code words are assigned to the messages. First generation approaches put the emphasis on the second step, whereas second generation approaches put it on the first step and use available results for the second step. Examples of first generation approaches are algorithms based on the discrete cosine transform (DCT) and discrete wavelet transform (DWT). An image captured by the camera sensor nodes is first transformed to a domain that is more suitable for processing. Quantisation is then carried out on the transformed image data. The image compression algorithm is then applied followed by entropy coding. This generates a bit stream that is ready for transmission. At the decoder side, the above processes are reversed which finally leads to the reconstruction of the original image. Second generation approaches require more computational processing compared to first generation approaches. Thus, researchers have focused on adapting and modifying first generation algorithms to lower computational and memory requirements while still maintaining the coding performance. Further discussions of using event compression in WMSNs are given in Chap. 6.

2.4.3.3 Event Fusion

WMSNs can employ multi-view approaches to further reduce the data. Each WCH receives data from several WMNs. The FOVs of WMNs may overlap, and the WCH can further reduce the data by performing event aggregation by consolidating the information from the WMNs. This requires solving what is known in computer vision as the correspondence problem. This is to find a set of points in one image which can be identified as the same points in another image. The application of multi-view compression approaches for WMSN is a very recent development. Note that the role of a WCH could be played by WMNs as well. For example, the WMN with the highest remaining energy level could be designated as the current WCH, and when its energy level decreases to a certain level, another WMN could take over the role of the WCH. Further discussions of using event fusion or stitching in WMSNs are given in Chap. 7.

2.5 Applications of WMSNs

The applications of WMSNs include current applications that make use of audio and visual sensors with perhaps the most common being video surveillance. Some applications of WMSNs are as follows:

- *Multimedia Surveillance Sensor Networks.* Multimedia surveillance sensor networks help enhance current surveillance systems by allowing large-scale deployment of low-cost low-powered audio-visual nodes. These nodes are capable of processing and transmitting multimedia data. Use of these networks can help deter crime through monitoring, detection and recording of potentially relevant activities (thefts, automobile collisions, traffic violations or suspicious behaviour). Audio can be used to detect abnormal sounds or disturbances such as gunshots, and a network of audio detectors can help determine the source of such occurrences.
- *Traffic Avoidance, Enforcement and Control Systems.* WMSNs can be used to monitor and determine the average travelling time on roads and highways. Data from this system can be used to offer traffic routing services to avoid congestion and to reduce user travelling time. An example of this is via GPS devices. The traffic data can be obtained via Radio Data System-Traffic Message Channel (RDS-TMC) or General Packet Radio Service (GPRS). WMSNs can also be deployed in intelligent parking systems to monitor available parking bays. This information can be used at an area wide level to help drivers decide where to park their vehicles (i.e. which parking facility), and once there the data can also be used for directing the drivers to the available parking bays.
- *Advanced Healthcare Delivery.* WSNs have been used to provide ubiquitous healthcare services [62, 148]. A patients vital and physical parameters such as pulse, blood pressure and body temperature can be monitored, recorded and transmitted to be stored for immediate or later access. WMSNs with the ability to deliver multimedia information will allow a greater wealth of information to be gathered.
- *Environmental and Home Monitoring.* WMSNs are highly suitable for habitat and home monitoring applications as they can be densely deployed for complete coverage. For habitat monitoring, WMSNs can aid in studying the patterns of wildlife in their natural environment. The inconspicuous WMSN nodes can provide a good coverage of the environment while not affecting the environment or causing changes to the behaviour of the subjects being monitored. Home monitoring is important for those in need of care such as the elderly and those who are physically or mentally impaired. Visual information from the network can be used to detect and infer if untoward occurrences have happened such as

medical emergencies, falls and even burglaries or thefts. The system can then contact emergency services or remote assistance services.

* *Industrial Process Control.* Visual information from the visible and non-visible spectrum can be used for industrial process control. This can be used for both quality control and plant monitoring. The use of imaging sensors for quality control is not new in quality control, but the WMSN will allow greater flexibility for its placement to have finer monitoring on the manufacturing process. The audio-visual sensors can also be used in the same manner as that in environment monitoring to detect and infer emergency or exceptional situations.

2.6 Chapter Summary

This chapter has presented an overview of existing technologies for WMSNs. These include the network architectures of WMSNs, node technologies and a discussion of some of the components and information processing in WMSNs. There are many design challenges arising from the limited resources in WMSNs such as size, cost and power. This chapter also presented a possible solution to the increase of data generated through the use of in-network data processing for redundancy elimination from multi-camera systems and selective transmission and compression of the data. The following chapters will describe these solutions in detail and how it can be achieved through the use of reconfigurable hardware.

Chapter 3
Hardware Technology and Programming Languages for Reconfigurable Devices

Abstract This chapter discusses the hardware technology and programming languages for reconfigurable devices. The first part of the chapter discusses the technology available and the advantages of field programmable gate arrays (FPGAs) for implementation in hardware constrained environments such as wireless multimedia sensor networks and describes the range of FPGA technology and internal architectures using examples from Xilinx and Altera FPGA families. The second part of the chapter presents an overview of hardware description languages for reconfigurable devices. Here, ranges of available languages from lower-level languages like VHDL and Verilog to higher-level C-based languages are discussed. The chapter also contains an introduction to the Handel-C programming language which will be used in the next chapter for the processor specification. The DK Design Suite methodology used by Celoxica is also discussed.

3.1 Introduction

There are currently three options a designer has to implement algorithms in hardware: hardwired logic or application-specific integrated circuits (ASICs), software programmed microprocessors and reconfigurable computing. The advantage of using ASICs is that they can be designed specifically to meet the requirements for a particular application to give fast and efficient computations. The disadvantage is that any modifications to the functionality would require the entire IC to be replaced which would be a costly process. Software programmed microprocessors provide flexibility by allowing changes to the software instructions to be executed. However, this comes at a cost of performance, and typically, microprocessor-based designs are slower than ASICs due to the high overheads required for processing each instruction. The third approach is to use reconfigurable hardware which provides higher performance than software programmed microprocessors while also maintaining a higher level of flexibility than ASICs. A reconfigurable computer has the facility to make significant changes to its datapath and control. The hardware can be specially configured to perform an application-specific task such as image processing or pattern matching, and after completion the hardware can be reconfigured to perform another specific task. The concepts behind reconfigurable computing can be traced back to Estrin in 1960 and the fixed plus variable (F+V) structure computer that described a computer system that consisted of a main processor augmented by an array of reconfigurable hardware [40]. The F+V machine dynamically adapted its computational resources for different algorithms to achieve high speed execution.

The work in [121] reports on an implementation of a point multiplication operation in elliptic curve cryptography using reconfigurable technology. For a key size of 270 bits, the reconfigurable hardware implemented on a Xilinx XC2V6000 field programmable gate arrays (FPGA) computed a point multiplication in 0.36 ms at a clock frequency of 66 MHz. In comparison, a software implementation implemented on a Intel dual Xeon processor required 196.71 ms at a clock frequency of 2.6 GHz. The reconfigurable hardware design gave a performance which was 540 times faster at a clock frequency 40 times lower than the Xeon processor. For implementation in embedded systems and hardware constrained devices such as wireless multimedia sensor networks (WMSNs), reconfigurable devices give further advantages in terms of reduced energy and power consumption. The work in [116] showed that average energy savings of 35–70 % with an average speed-up of three to seven times were achieved when critical software loop instructions were implemented in reconfigurable hardware. Other advantages of reconfigurable hardware include a reduction in size and component count, faster time to market, improved flexibility and upgradability [123].

As integrated circuits grew in complexity, a better method for designing them was needed. One approach is to use schematic capture tools to draw schematics on a computer screen to represent a circuit. This worked well because graphical representations are useful to understand small but complex functions. However, as chip density increased, schematic capture of these circuits became difficult to use for the design of large circuits. The transistor densities of ASICs and FPGAs grew to a point where better tools were needed. The development of hardware description languages (HDLs) are tools that are being widely used today in many electronic designs. A HDL is a language from a class of computer languages, specification languages or modelling languages for formal description and design of electronic circuits. The HDL can be used to describe the circuit operation, its design, organisation and testing to verify its operation by means of simulations. There are currently many HDLs which have been developed to target different levels and abstractions of the electronic design process.

3.2 Hardware Technology for Reconfigurable Hardware

This section will discuss technology for reconfigurable hardware in terms of the system-level architectures, reconfigurable-level models and functional block architectures. The section will also give an overview of soft-core microprocessors such as the Xilinx MicroBlaze and Altera Nios.

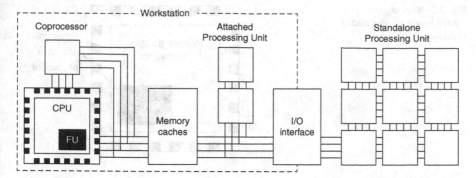

Fig. 3.1 Reconfigurable configurations for external processing unit, attached processing unit, co-processor and reconfigurable functional unit

3.2.1 System-Level Architectures

The system-level architectures for reconfigurable systems can be classified into five configurations depending on the degree of coupling between the reconfigurable hardware and the host processor (CPU) [123]. The five configurations are standalone processing unit, attached processing unit, co-processor unit, reconfigurable functional unit and embedded processor unit. Figure 3.1 shows the configurations for the standalone processing unit, attached processing unit, co-processor unit and reconfigurable functional unit.

3.2.1.1 Standalone Reconfigurable Processing Unit

The standalone or external processing unit configuration is the most loosely coupled form of architecture for reconfigurable hardware. The reconfigurable hardware does not communicate frequently with the processor, and data communication transfer is slow. The processor and reconfigurable hardware are often placed on different chips. This architecture is suitable for applications where a significant amount of the processing is executed on the reconfigurable hardware without needing the processor to intervene. This configuration is often used by emulation software.

3.2.1.2 Attached Reconfigurable Processing Unit

The configurations for the attached processing unit and co-processor unit have a tighter coupling than the standalone processing unit configuration. For the attached processing unit configuration, the reconfigurable hardware is placed on the same chip as the processor and attached to the processor bus. The data cache for the host processor is not visible to the reconfigurable hardware leading to a high delay

Fig. 3.2 Reconfigurable configuration for embedded processor

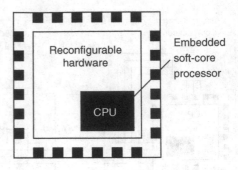

in communication between the processor and the reconfigurable hardware. The advantage of the configuration is that it does allow for a high amount of computation independence by shifting over to the reconfigurable hardware. This configuration can be used as an additional processor in a multiprocessor system.

3.2.1.3   Reconfigurable Co-processor Unit

A co-processor is able to perform computations independently of the host processor. After the co-processor configuration has been initialised, the reconfigurable hardware can operate without processor supervision and return the result after completion. This also allows the host processor and the reconfigurable hardware to perform their executions at the same time.

3.2.1.4   Reconfigurable Functional Unit

For the reconfigurable functional unit configuration, the reconfigurable hardware forms part of the processor itself and is used to implement application-specific instructions. The reconfigurable units execute as functional units on the processor datapath using registers to hold the input and output operands. The customised instructions implemented in the reconfigurable hardware are typically reconfigured between applications, and a customised instruction set can be selected based on performance improvement for a specific application.

3.2.1.5   Reconfigurable Embedded Processor Unit

Figure 3.2 shows the configuration for the embedded processor architecture. In this configuration, the processor is embedded in the reconfigurable hardware. The processor may even be a soft-core processor which is implemented using the resources of the reconfigurable hardware itself. Examples of soft-core processors are the Xilinx MicroBlaze and the Altera Nios.

3.2.2 Reconfigurable-Level Models

Reconfigurable-level models can be classified as either mask programmable or field programmable. Early reconfigurable devices were mask programmable which was performed by the manufacturer with configuration information from the customer. Later on, manufacturers allowed the devices to be reconfigured after fabrication or in the field. This was termed as field programmable devices of which FPGAs are the most well known. Reconfigurable-level models can also be classified as statically reconfigurable or dynamic reconfigurable.

3.2.2.1 Static Reconfiguration Models

Static reconfiguration or also known as compile-time reconfiguration is the simplest and most common model for implementing applications with reconfigurable logic. The reconfiguration process is slow, and the application has to be halted when the reconfiguration process is being carried out. After the reconfiguration, the application is restarted with the new program.

3.2.2.2 Dynamic Reconfiguration Models

Dynamic reconfiguration or also known as run-time reconfiguration allows for the hardware to be loaded and unloaded dynamically during the operation of the system. This model of reconfiguration also allows for the reconfigurable hardware to support more functionality than could be supported by the physical hardware resources itself. The application functionality is partitioned into several configurations, and each configuration is loaded and unloaded dynamically as required. For example, a speech coding application could begin with a filtering configuration. After the filtering process has been completed, the configuration is swapped or loaded out and a coding configuration swapped into the hardware to perform the coding process. Four main categories can be distinguished in terms of the underlying mechanism to reconfigure the devices [33]. The categories are single context, multi-context, partial reconfigurable and pipeline reconfigurable.

Single-Context Reconfigurable Model

A single-context reconfigurable architecture has a single configuration memory bit to hold each configuration data bit. Any change to a single-context configuration would require a complete programming of the entire chip. Consequently, this architecture would incur a high overhead cost if only a small part of the configuration memory needs to be changed. Furthermore, the system execution needs to be halted during the reconfiguration. This model is more suitable for architectures that do not require run-time reconfiguration. To implement run-time reconfiguration on a

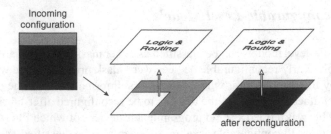

Fig. 3.3 Single-context reconfigurable model

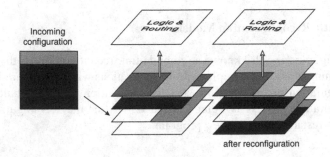

Fig. 3.4 Multi-context reconfigurable model

single-context FPGA, the configurations are grouped into contexts, and each full context is swapped in and out of the FPGA as required. Examples of single-context FPGAs are the Xilinx 4000 series and the Altera Flex10K series. Figure 3.3 shows the single-context reconfigurable model.

Multi-context Reconfigurable Model

A multi-context FPGA includes multiple memory bits for each programming bit location. This reconfigurable architecture can switch very fast between different configurations, and new configurations can be downloaded in parallel with an actively used configuration. A multi-context device can be considered as a multiplexed set of single-context devices. A particular context is selected and fully programmed to perform any modification to the device. Figure 3.4 shows the multi-context reconfigurable model.

Partially Reconfigurable Model

In a partially reconfigurable FPGA, the entire FPGA does not have to be reprogrammed to load in a new configuration. The underlying programming bit layer operates like a RAM device and allows only the selected target addresses to be modified for selective reconfiguration. This also has the benefit of hiding the

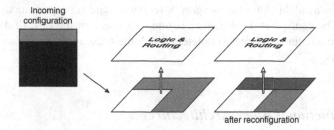

Fig. 3.5 Partial reconfigurable model

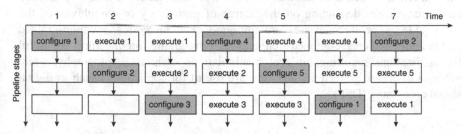

Fig. 3.6 Pipeline reconfiguration model

reconfiguration latency as the non-changed part of the reconfigurable hardware can continue normal execution at the same time that the reconfiguration is being performed. A practical example used by Xilinx is to divide the FPGA board into two different areas: static region and partially reconfigurable region. The partially reconfigurable region can be reconfigured while mission-critical operations in the static region continue to perform without interruption. To support partial reconfiguration, address information has to be supplied together with the configuration data, and the total information transferred would be greater than for the single-context model. This model is suitable for applications where the size of the configurations is small enough to fit several configurations on the hardware at the same time. Figure 3.5 shows the partial reconfiguration model.

Pipeline Reconfigurable Model

The pipeline-reconfigurable model [18] is a specific case of the partially-reconfigurable model where the reconfiguration occurs in increments of pipeline stages. This model is suitable for datapath computations where the entire pipeline datapath does not fit on the available physical hardware. Each pipeline stage is mapped as a configuration and configured as a whole. These models allow for the overlap of configuration and execution. A pipeline stage is configured as the other pipeline stages are in execution. Figure 3.6 shows the pipeline reconfigurable model for three physical pipeline stages implementing five virtual pipeline stages. The pipeline stages are configured one by one, from the start of the pipeline through the end of the available hardware stages. After each stage is programmed, it begins execution. The configuration of a stage is one step ahead of the flow of data. After the hardware pipeline has been completely filled, the reuse of the hardware pipeline stages begins.

3.2.3 Functional Block Architectures

The functional block architectures for reconfigurable hardware can be classified into three classes depending on the degree of granularity or complexity of the block [118]. The three classes are fine-grained architectures, medium-grained architectures and coarse-grained architectures. Fine-grained architectures are useful for applications requiring bit-level manipulations such as cryptography, while medium to coarse-grained architectures are useful for applications such as digital signal processing (DSP).

3.2.3.1 Fine-Grained Architecture

The FPGA is an example using a fine-grained architecture. This section will describe the basic architecture of an FPGA and give example architectures from Xilinx and Altera.

Field Programmable Gate Array

The FPGA is a reconfigurable device that contains programmable functional blocks, programmable interconnects (connection blocks) and switches. Figure 3.7 shows an overview of an FPGA architecture. The functional blocks in an FPGA operate at a bit-level and allow it to implement datapaths and controllers with different word lengths. In commercial FPGAs, the functional blocks contain small lookup tables (LUTs) which have up to six inputs to implement the hardware logic. The block would also contain other elements such as registers, multiplexor and I/O ports to the programmable interconnects. FPGA manufacturers use different terms to describe their functional blocks. Xilinx uses the term configurable logic blocks (CLBs), whereas Altera uses the term logic array blocks (LABs). The programmable interconnects connect the signals from the functional blocks, and the switches allow the signals to change direction in the horizontal and vertical channels. Next, we will briefly describe the internal architectures for two FPGAs from Xilinx and Altera which are the Spartan-3 FPGA and the Stratix II FPGA.

Figure 3.8 shows the architecture of a Xilinx Spartan-3 FPGA [140]. The architecture consists of five basic element block types: CLBs, input/output blocks (IOBs), block RAMs (BRAMs), multiplier blocks and digital clock manager (DCM) blocks. The CLBs are grouped into rows and columns across the device. The CLBs

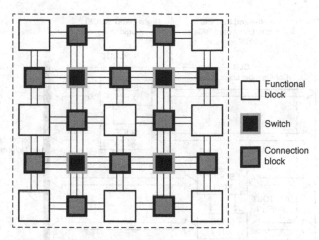

Fig. 3.7 Overview of an FPGA architecture

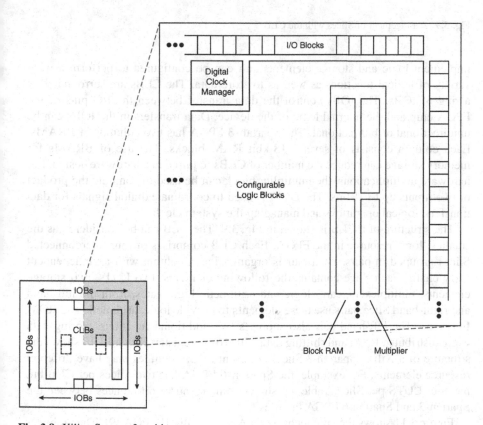

Fig. 3.8 Xilinx Spartan-3 architecture

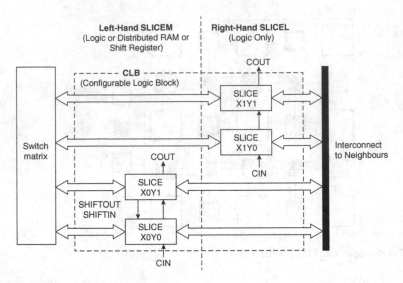

Fig. 3.9 Arrangement of Slices with the CLB

implement logic and storage elements and can be configured to perform a wide variety of logical functions as well as to store data. The CLBs are surrounded by a ring of IOBs. The IOBs control the data transfer between the I/O pins of the FPGA chip and the internal logic of the device. Data transfers in the IOBs can be unidirectional or bidirectional. The Spartan-3 FPGA has two columns of BRAMs. Each column consists of several 18 kbit RAM blocks. The use of BRAMs for memory storage can reduce the number of CLBs required in a hardware design. For hardware multiplications, the multiplier block can be used to compute the product of two inputs up to 18 bits. The DCM is used to coordinate digital signals for data transfers, logical operations and managing the system clock.

The structure of a CLB is shown in Fig. 3.9. The CLB can be considered as the smallest logic resource in the FPGA. Each CLB comprises of four interconnected Slices grouped in pairs. Each pair is organised as a column with an independent carry chain. Each Slice contains the following elements: two LUTs, two storage elements, multiplexers, carry logic and arithmetic logic gates. Both the left-hand and right-hand Slice pairs use these elements to provide logic, arithmetic and ROM functions. The left-hand pair also supports two additional functions: storing data using distributed RAM and shifting data with 16-bit registers. Figure 3.10 shows the structure of a Xilinx Spartan-3 Slice. Different FPGA families may have different resource elements. For example, the Spartan-6 FPGA has two Slices per CLB but has four LUTS per Slice. Table 3.1 shows some resource differences between the Spartan-3 and Spartan-6 FPGA families.

Figure 3.11 shows the architecture of a Altera Stratix II FPGA [9]. The architecture consists of three basic element block types: LABs, memory block structures and DSP blocks. The LABs are grouped into rows and columns across the device.

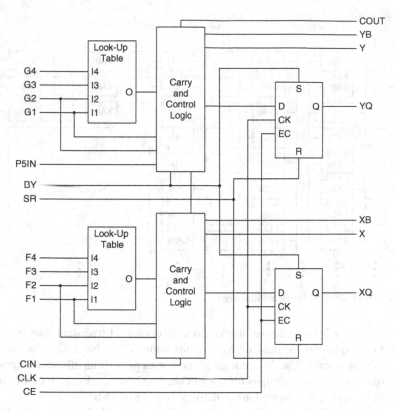

Fig. 3.10 Xilinx Spartan-3 slice structure

Table 3.1 Resource differences between Spartan-3 FPGA and Spartan-6 FPGA

Resource elements	Spartan-3	Spartan-6
Lookup tables (LUTs)	4-input	6-input
Slices per CLB	4	2
LUTs per slice	2	4

Each LAB consists of eight adaptive logic modules (ALMs). An ALM is the basic building block for the Stratix II device family providing efficient implementation of logic functions. The memory block structures have three types of memory blocks: M512 RAM, M4K RAM and M-RAM. The M512 RAM blocks are simple dual-port memory blocks with 512 bits plus parity (576 bits). These blocks provide dedicated simple dual-port or single-port memory up to 18-bits wide. These blocks are grouped into columns across the device in between certain LABs. The M4K RAM blocks are true dual-port memory blocks with 4k bits plus parity (4,608 bits). These blocks provide dedicated true dual-port, simple dual-port or single-port memory up to 36-bits wide. These blocks are grouped into columns across the device in between certain LABs. The M-RAM blocks are true dual-port memory blocks with 512k bits

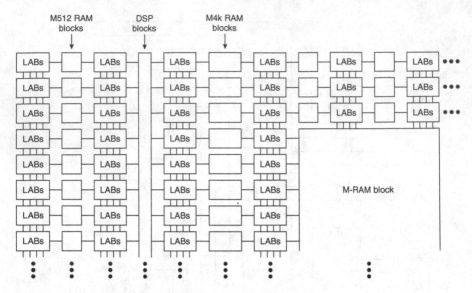

Fig. 3.11 Altera Stratix II FPGA architecture

plus parity (589,824 bits). These blocks provide dedicated true dual-port, simple dual-port or single-port memory up to 144-bits wide. These blocks are located in the device's logic array. The different sized memories provide an efficient mapping for variable-sized memory designs for the FPGA. The DSP block can be configured to support a single 36×36 bit multiplication, four 18×18 bit multiplications or eight 9×9 bit multiplications. These blocks also contain shift registers for DSP applications, including finite impulse response (FIR) and infinite impulse response (IIR) filters. The DSP blocks are grouped into columns across the device.

The structure of an ALM is shown in Fig. 3.12. The ALM is able to adaptively adjust the effective size of the LUTs to better use the logic resources. Each ALM contains a variety of look-up table (LUT)-based resources that can be divided between two adaptive LUTs (ALUTs). With up to eight inputs to the two ALUTs, one ALM can implement various combinations of two functions. The ALM can be combined as two independent 4-LUTs, a 5-LUT and 3-LUT, a single 6-LUT, two independent 5-LUTs with two inputs shared between the LUTs or two 6-LUTs with the same function with four inputs shared between the LUTs. In addition to the adaptive LUT-based resources, each ALM contains two programmable registers, two full adders, a carry chain, a shared arithmetic chain and a register chain.

3.2.3.2 Coarse-Grained Architecture

Examples of medium-grained to coarse-grained architectures are ALU arrays and processor arrays. Medium-grained architectures use more than a single bit to implement the functional blocks. An example is the Garp architecture [59] which uses 2-bit wide functional blocks.

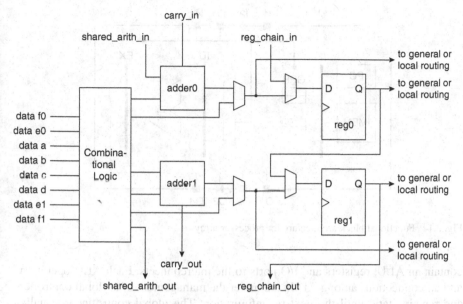

Fig. 3.12 Altera Stratix II ALM structure

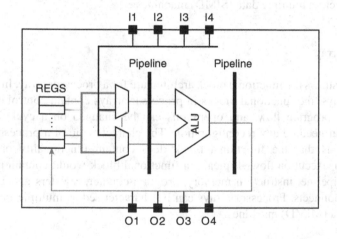

Fig. 3.13 Functional block architecture for ALU array

ALU Array

Figure 3.13 shows the functional block architecture for an ALU array. Each functional block consists of an ALU that supports word-level operations. The characteristic of an ALU array architecture is that the functional block does not contain any functionality or control elements for execution flow such as an instruction memory or execution pipeline. Typically a functional block would only

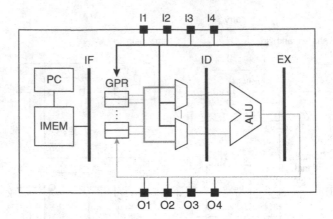

Fig. 3.14 Functional block architecture for processor array

contain an ALU, registers and I/O ports to the interconnects. Each ALU operation and all connections among ALUs are under the management of a global controller and remain static until the next reconfiguration. The global controller is usually implemented as a separate attached processor. ALU arrays can be characterised as single instruction multiple data (SIMD) machines.

Processor Array

Figure 3.14 shows the functional block architecture for a processor array. In contrast to ALU arrays, the functional blocks in processor arrays contain control elements for its own program flow, and operations can be changed on a cycle-by-cycle basis without needing any reconfiguration. The characteristic of a processor array architecture is that the functional block does contain functionality or control elements for execution flow. Typically a functional block would contain an ALU, execution pipeline, instruction memory, program counter, registers and I/O ports to the interconnects. Processor arrays can be characterised as multiple instruction multiple data (MIMD) machines.

3.2.4   Soft-Core Microprocessor Architectures

A soft-core processor is a microprocessor described using a HDL which can be synthesised and implemented in reconfigurable hardware such as FPGAs. Using a soft-core processor design over a traditional hard-core or general-purpose processor design gives two advantages. The first advantage is the flexibility it provides because its parameters can be changed by reprogramming the device. The second advantage is that the design is platform independent and can be synthesised for any

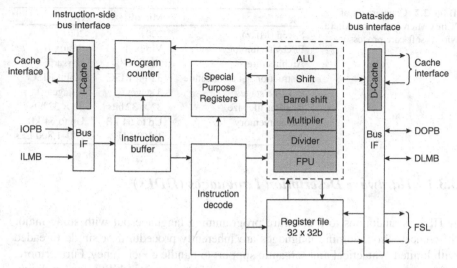

Fig. 3.15 Xilinx MicroBlaze soft-core processor architecture

technology including FPGAs and ASICs. Two examples for soft-core processors
are the Xilinx MicroBlaze [139] and the Altera Nios II [8]. The MicroBlaze is
a 32-bit processor that is optimised for embedded systems applications. It uses a
Harvard RISC architecture with a three-stage pipeline and a 32 × 32 bit register
file. The processor can be configured with an optional IEEE-754 floating-point unit,
hardware divider, barrel shifter and exception handling features. The MicroBlaze is
targeted for implementation on the Virtex and Spartan FPGA families. Figure 3.15
shows the Xilinx MicroBlaze soft-core processor architecture. The Nios II also uses
a Harvard RISC architecture with a 32 bit instruction set architecture (ISA). It can
be configured with hardware logic for single-instruction 32 × 32 multiply and divide
instructions or 64 and 128 bit multiply instructions. The soft-core processor can be
configured in three configurations using an economy core, standard core or fast
core depending on the application requirements. The economy core is optimised for
area and is suitable for small embedded applications. The fast core is optimised
for performance for computationally intensive applications such as multimedia
processing. Table 3.2 shows a comparison of the Xilinx MicroBlaze and the Altera
Nios II soft-core processors.

3.3   Programming Languages for Reconfigurable Devices

This section will discuss HDLs for reconfigurable hardware in terms of the HDL
design flow and abstraction models such as behavioural models and structural
models. The section will also give an overview of lower-level HDLs like VHDL
and Verilog and higher-level HDLs like SystemC, SpecC and Handel-C.

Table 3.2 Comparison of Xilinx MicroBlaze and Altera Nios II soft-core processors

	MicroBlaze	Nios II
Speed (MHz)	200	200
FPGA technology	Virtex-4	Stratix II
Architecture	Harvard	Harvard
Instruction set architecture	32-bit RISC	32-bit RISC
Pipeline stages	3 stages	6 stages
Register file size	32×32 bits	32×32 bits
Cache memory	Up to 64 kB	Up to 64 kB
Area	1,269 LUTs	700–1,800 LEs

3.3.1   Hardware Description Languages (HDLs)

A HDL is analogous to a software programming language but with some major differences. Programming languages are inherently procedural or single threaded with limited syntactical and semantic support to handle concurrency. Furthermore, software programming languages do not include any capability for explicitly expressing time. HDLs, on the other hand, are standard text-based expressions of the spatial and temporal structure and behaviour of electronic systems. As its syntax and semantics can include explicit notations for expressing concurrency, the HDL can model multiple parallel processes such as flip-flops and adders that execute independently of one another. HDLs also include an explicit notion of time which is a primary attribute of hardware. Any changes to the input into a hardware process can automatically trigger an update in the process stack of the simulator. HDLs serve two purposes: simulation and synthesis. First, it is used to write a model for the expected behaviour of a circuit before that circuit is designed and built. The model is fed into a computer program called a simulator which allows the designer to verify that the solution behaves correctly. This process is known as hardware simulation. Second, it is used to write a detailed description of a circuit that is input into another computer program called a logic compiler. The output of the compiler is used to configure a programmable logic device such as an FPGA to perform the hardware function. This process is known as hardware synthesis. The processes of simulation and synthesis will be further discussed in the next section.

3.3.2   HDL Design Flow

Figure 3.16 shows a general design flow using HDL. The two major types of HDL processing are the simulation process and the synthesis process. The simulation process is the interpretation of the HDL statements for the purpose of producing human readable output such as a timing diagram that predicts how the hardware will behave before it is actually fabricated. Simulation is useful to a designer because it allows detection of functional errors in a design without having to fabricate the actual hardware. When a designer catches an error with simulation, the error

Fig. 3.16 General design
flow using hardware
description language

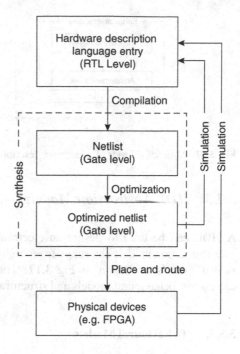

can be corrected with a few changes to the HDL code. If the error is not caught
until the hardware has been fabricated, correcting the problem would be much
more costly and complicated. The synthesis process involves the compilation of
high-level behavioural and structural HDL statements into a flattened gate-level
netlist. This netlist can then be used directly either to lay out a printed circuit board,
to fabricate a custom integrated circuit or to program a programmable logic device
such as an FPGA. The synthesis process produces a physical piece of hardware that
implements the computation described by the HDL code.

The synthesis process can be divided into three stages: compilation, optimisation
and place and route stages. Compilation is the conversion of the high-level HDL
language which describes the circuit at the register-transfer level (RTL) into a
netlist at the gate level. Following this, optimisation is then performed on the
gate-level netlist. The optimisations can be in terms of speed to have a fast piece of
hardware or in terms of area to minimise the hardware resources requirements. The
place-and-route process comes after the generation of a netlist and is composed of
two steps: placement and routing. The first step is placement and involves deciding
where to place the electronic components, circuitry and logic elements on the
hardware space. This is followed by routing which decides the exact design for
all the wires needed to connect the placed components. This step implements all the
desired connections while following the rules and limitations of the manufacturing
process. Finally, the physical layout is generated for an FPGA chip, or the mask is
generated for an ASIC.

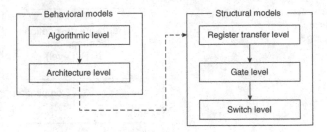

Fig. 3.17 Abstraction levels of a hardware description language

3.3.3 HDL Abstraction Models

A HDL can be used to design an electronic system at any level of abstraction. The abstraction models can range from high-level architectural models to low-level switch models as shown in Fig. 3.17. These levels can be classified into two categories: behavioural models and structural models.

3.3.3.1 Behavioural Models

Behavioural models consist of code that represents the behaviour of the hardware without respect to its actual implementation. Behavioural models provide a high level of abstraction and do not include timing information. For example, hardware adders can be specified to add two or more numbers without having to specify any register, gate or transistor details. Also, the specification of hardware buses does not need to be broken down into their individual signals. Behavioural models can be classified further as algorithmic or architectural models. Algorithms are step-by-step methods of solving problems and producing results. No hardware implementation is implied in an algorithmic model. An algorithmic model is not concerned with clock signals, reset signals or any kind of signals. It does not deal with flip-flops, gates or transistors. There is no need to specify how a numeric quantity is represented in hardware or what kind of memory device is used to store it. Architectural models describe hardware at a very high level using functional blocks like memory, control logic, central processing unit or first-in-first-out buffers. These blocks may represent personal computing boards such as ASICs, FPGAs or other major hardware components of the system. An architectural model deals with hardware issues like how to represent signed integers or how much memory is required to store intermediate values. An architectural description may involve clock signals but is usually not to the extent that it describes everything that happens in the hardware on every clock cycle. That is performed by a RTL description. In general, an architectural description is useful for determining the major functions of a design.

3.3.3.2 Structural Models

Structural models consist of code that represents specific pieces of hardware. The design approach follows a hierarchical model. The lowest level in the hierarchy is composed of primitives provided by the specific language being used. Structural models can be further classified as RTL, gate level or switch level models. RTL models specify hardware on a register level and what happens on each clock edge. Actual gate descriptions are usually avoided, although RTL code may use Boolean functions that can be implemented in gates. The high-level descriptions of hardware functions can be used as long as these functions have a behaviour that is associated with clock cycles. An example of an RTL description is a state machine. The functionality of a state machine can be complex, but the RTL state machine is defined by what occurs on each clock cycle. Gate-level models consist of code that specifies very simple functions such as NAND and NOR gates. The lowest level description is that of switch-level models which specifies the actual transistor switches that are combined to make gates. ASIC vendors and semiconductor manufacturers may use this level of modelling to simulate their circuits.

3.3.4 Advantages of HDLs

Electronic designs described in HDLs are technology independent, easy to design and debug and are usually more readable than schematics, particularly for large circuits. The advantages of using HDLs include [143]:

- *Reusability*. Reusability is a big advantage of HDLs. The code written in one HDL can be used in any system that supports that HDL. Schematics are only useful in a particular schematic capture software tool. Even using the same tool, portability can be difficult if a module does not physically fit into the new design. A behavioural level HDL model and most RTL level models can be reused for multiple designs.
- *Concurrency*. Concurrency is an advantage that HDLs offer over normal software languages to simulate hardware. With a software language, statements are executed sequentially. Provisions have been added to HDLs to support concurrent execution of statements. This is a necessity since in a hardware design, many events occur simultaneously. For example, in a synchronous design, all flip-flops on a particular clock line must be evaluated simultaneously. While normal software languages can be used to model simultaneous events, it is the responsibility of the programmer to add the mechanism for handling this. For HDLs, the mechanism is built into the language.
- *Timing*. With schematic capture tools, when it comes for the time to simulate a design, the timing numbers are embedded in the netlist that is generated from the schematic. These timing numbers are based on parameters supplied by the vendors whose chips are being used. The user has some limited ability to change these numbers and some of the parameters. With HDLs, the timing numbers are

explicitly stated in the code. Nothing is hidden from the user, and the user has complete control over these numbers. This makes it easier to control and optimise the timing of the design.

- *Optimisation*. HDLs are particularly useful for optimisation. One common method of designing ASICs and FPGAs involves writing an RTL level design and then using a software tool to synthesise the design. The synthesis process involves generating an equivalent gate-level HDL description which can then be optimised to suit the particular target hardware. An RTL level model can be written and synthesised specifically for a particular FPGA. The same RTL description can be used and synthesised and optimised for a particular ASIC technology.
- *Standards*. Major HDLs such as Verilog and VHDL are public standards. Verilog has been adopted by the Institute of Electrical and Electronic Engineers (IEEE) as the standard IEEE-STD-1364. VHDL has been adopted by the IEEE as the standard IEEE-STD-1076. Because these languages are public standards, it ensures that a design written in the language can be accepted by every software tool that supports the language and a designer has a wide range of tools to be selected from many vendors.
- *Documentation*. HDLs are text-based programming-type languages and lend themselves easily to documentation. HDL code is relatively easy to read, and a code module shows much about the functionality, the architecture and the timing of the hardware. Furthermore, statements can be organised in ways that give more information about the design, and comments can be included which explain the various sections of code. The nature of HDLs encourages this type of code documentation.

3.3.5 Examples of HDLs

HDLs can be classified into two categories: lower-level HDLs like VHDL and Verilog and higher-level HDLs like SystemC, SpecC and Handel-C. This section will give a discussion on these HDLs.

3.3.5.1 Lower-Level HDLs

Examples of lower-level HDLs are the very high-speed integrated circuit hardware description language (VHDL) and Verilog.

VHDL

VHDL [63] was developed and supported by the United States Department of Defence. It is a very powerful HDL used in electronic design automation to describe digital and mixed-signal systems such as field-programmable gate arrays

Table 3.3 VHDL built-in data operators

Symbol	Operation	Example
Logical operator		
AND	AND	n <= a AND b
OR	OR	n <= a OR b
NOT	NOT	n <= a NOT b
NAND	NAND	n <= a NAND b
NOR	NOR	n <= a NOR b
XOR	XOR	n <= a XOR b
XNOR	XNOR	n <= a XNOR b
Arithmetic operator		
+	Addition	n <= a	b
−	Subtraction	n <= a − b
*	Multiplication	n <= a * b
/	Division	n <= a / b
MOD	Modulus	n <= a MOD b
REM	Remainder	n <= a REM b
**	Exponentiation	n <= a ** b
&	Concatenation	n <= a & b
ABS	Absolute	n <= ABS(a)
Relational operator		
=	Equal	IF (n = 10) THEN
/=	Not equal	IF (n/= 10) THEN
<	Less than	IF (n < 10) THEN
<=	Less than or equal	IF (n <= 10) THEN
>	Greater than	IF (n > 10) THEN
>=	Greater than or equal	IF (n >= 10) THEN
Shift operator		
SLL	Shift left logical	n <="1001010" SLL 2
SRL	Shift right logical	n <="1001010" SRL 1
SLA	Shift left arithmetic	n <="1001010" SLA 2
SRA	Shift right arithmetic	n <="1001010" SRA 1
ROL	Rotate left	n <="1001010" ROL 2
ROR	Rotate right	n <="1001010" ROR 2

and integrated circuits. VHDL allows for the description of a digital system at the
structural or the behavioural level. VHDL was developed to model all levels of
design. VHDL can accommodate behavioural constructs and mathematical routines
that describe the complex models. The behavioural level can be further divided into
two kinds of approaches: dataflow and algorithmic. The dataflow representation
describes how data moves through the system. This is typically done in terms of
dataflow between registers which is also termed as the RTL. The data flow model
makes use of concurrent statements that are executed in parallel as soon as data
arrives at the input. The algorithmic approach consists of sequential statements that
are executed in the sequence that they are specified. Table 3.3 shows the built-in data

Table 3.4 Verilog built-in data operators

Symbol	Operation	Example
Logical Operator		
&&	AND	n <= a && b
||	OR	n <= a || b
!	NOT	n <=~ a
&	Bitwise AND	n <= a & b
|	Bitwise OR	n <= a | b
~	Bitwise NOT	n <= a ~ b
^	Bitwise exclusive OR	n <= a ^ b
Arithmetic Operator		
+	Addition	n <= a + b
−	Subtraction/Negation	n <= a − b
*	Multiplication	n <= a * b
/	Division	n <= a / b
%	Remainder	n <= a % b
Relational Operator		
==	Equal	IF (n == 10) THEN
! =	Not equal	IF (n! = 10) THEN
<	Less than	IF (n < 10) THEN
<=	Less than or equal	IF (n <= 10) THEN
>	Greater than	IF (n > 10) THEN
>=	Greater than or equal	IF (n >= 10) THEN

operators for VHDL. It includes logical operators, arithmetic operators, relational operators and shift operators.

Verilog

Similar to VHDL, the Verilog HDL [13] allows an engineer to design a digital design at behaviour level, RTL, gate level or at switch level. Table 3.4 shows the built-in data operators for Verilog. Although Verilog and VHDL does have a similar programming capability in hardware design, they do have some differences. For example, VHDL allows user defined types whereas in Verilog, only predefined data types are allowed. Also VHDL uses libraries to define entities, architectures, packages and configurations. On the other hand, Verilog has no library concept. This has caused it to be not very useful for multiple design projects. Another difference is that test harnesses are not supported in Verilog whereas VHDL is supported by generic and configuration statements.

3.3.5.2   Higher-Level and C-Based HDLs

The emergence of the system-on-chip (SoC) era is creating many new challenges at all stages of the design process. As electronic systems grow increasingly complex and reconfigurable systems become increasingly mainstream, problems arising from the use of different design languages, incompatible tools and fragmented tool flows are becoming common. Systems and software engineers do programming in C/C++, and their hardware counterparts use HDLs such as VHDL and Verilog. There is a growing desire in the industry for a single language that can perform some tasks of both hardware design and software programming. This section describes three modelling languages based on C that addresses this issue: SystemC, SpecC and Handel-C.

SystemC

SystemC [84] is a system-level modelling language based on C++ that is intended to provide a solution for representing functionality, communication, software and hardware at various levels of abstraction. It combines high-level languages with concurrency models to allow faster design cycles for FPGAs than is possible using traditional HDLs. SystemC inherits the features of the C++ class library and methodology. It provides an event-driven simulation kernel in C++ that can be used to create a cycle-accurate model of a system consisting of software, hardware and their interfaces. Similar to traditional HDLs, users can construct structural designs in SystemC using modules, ports and signals. Modules can be instantiated within other modules, enabling structural design hierarchies to be built. Ports and signals enable communication of data between modules, and all ports and signals are declared by the user to have a specific data type. SystemC supports the use of special data types which are often used by hardware engineers. The data types it supports include single bits, bit vectors, characters, integers, floating-point numbers and vectors of integers. It also has support for four-state logic signals (i.e. signals that model 0, 1, X and Z). In addition, an important data type that is found in SystemC but not in traditional HDLs is the fixed-point numeric type. Fixed-point numbers are frequently used in DSP applications that target both hardware and software implementations. SystemC supports concurrency. To simulate the concurrent behaviour of the digital hardware, the simulation kernel is designed so that all the processes are executed concurrently irrespective of the order in which they are called.

SystemC provides three basic computation units: function, process and module. A function is defined in the same way as that in the C language. Processes are the basic behavioural entities of SystemC. Although, a process can contain function calls, it cannot invoke other processes. Therefore, hierarchical modelling of processes is not possible. Modules are structural entities which serve as basic blocks for partitioning a design. Modelling using modules reflects the structural hierarchy. The modules can be classified into two categories: leaf module and composite module. A leaf module contains processes which specify the functionality of the module but which does not contain any module. A composite module consists of the instantiation of other modules. Data transfer is modelled by connecting module ports through either signals or channels. SystemC only supports dynamic scheduling of execution sequence. There are two mechanisms for dynamic scheduling: static sensitivity and dynamic sensitivity. When designers use a static sensitivity mechanism, a list of signals is specified in a sensitivity list of a process. If the value of any signal in the sensitivity list of a process changes, the process starts or resumes execution. The dynamic sensitivity mechanism uses event-wait-notify to schedule processes. A process can wait and notify an event. If the waited event of a process is notified, the process starts or resumes execution. However, ports of modules cannot be connected through an event. Therefore, the event-wait-notify cannot be used for synchronisation between processes in different modules. Designers have to encapsulate events into a channel in order to achieve synchronisation between such processes.

SpecC

Similar to SystemC, the SpecC [84] language is also intended for specification and design of SoCs or embedded systems including software and hardware whether using fixed platforms, integrating systems from different intellectual properties (IPs) or synthesising the system blocks from programming or HDLs. SpecC is a system-level design language (SLDL) and is an extension of the ANSI C programming language. It supports modelling at all levels of abstraction, from purely functional models to cycle-accurate RTL models. In addition, SpecC models can be simulated, so that functionality and timing constraints of a design can be validated. SpecC also has a profiler with a user-friendly graphical interface. SpecC has par and pipe constructs for modelling parallelism and pipelined execution, respectively. This gives the designers more options for making implementation decisions. In addition, SpecC allows validation by simulation and validation by refinement. The correctness of the outcome models can be ensured by validating the initial models and refinement algorithms.

SpecC provides two basic computation units: function and behaviour. A SpecC function follows the same semantics and syntax as a function in the C language. A function can be called hierarchically and executes sequentially according to the calling sequence. The SpecC behaviour is specified by a behaviour definition. There are two types of behaviours: leaf behaviour and composite behaviour. A leaf behaviour may contain hierarchically called functions, but it does not contain any sub-behaviour instances, whereas a composite behaviour consists of sub-behaviours instances. SpecC provides two mechanisms to model execution sequence: static scheduling and dynamic scheduling. In the case of static scheduling, the sequence of execution of behaviours is explicitly specified with par, pipe and fsm constructs [144]. For dynamic scheduling, it uses event-wait-notify to schedule behaviours dynamically. SpecC has a data type event, and the wait and notify statements are used for synchronisation between behaviours. When the wait statement such as wait(e) is executed, the behaviour ceases to execute until the waited event e of a behaviour is notified with a notify e statement. Although SpecC and SystemC share many features, they have some main differences which would

Table 3.5 Comparison of SystemC and SpecC in terms of design modelling

Abstract models	Model aspect	SystemC	SpecC
Specification models	Functional block	Module	Behaviour
	Schedule	Event, signal	Event, definition(par)
	Data transfer	Signal	Variable
IP-assembly model	Structure blocks	Module	Behaviour
	Functional blocks	Process	Behaviour
	Schedule inside PEs	Event, signal	Event, definition(par)
	Schedule between PEs	Channel	Channel
	Data transfer inside PEs	Signal	Variable
	Data transfer between PEs	Channel	Channel
Bus-arbitration model		Same as Arch model	Same as Arch model
Bus-functional model		Same as Arch model	Same as Arch model
Implementation Model		Switch (SC_THREAD), SC_CTHREAD	Fsmd
		Function/module	Function/behaviour
		Signal	Buffered signal
		Bit	Bit
		Signal	Signal

be encountered when specification modelling is considered. Table 3.5 shows a comparison between SpecC and SystemC in terms of design modelling.

First, SpecC supports static scheduling using par, pipe and fsm constructs or with the default sequential execution. The static schedule allows designers to determine the explicitly modelled execution sequence which is used during architecture exploration. Static scheduling features are not available in SystemC. Second, SystemC uses module as the structural entity and process as the behavioural entity. It does not support hierarchical modelling of processes. Therefore, both process and module do not fully support behaviour entity modelling. SpecC does support modelling of the behavioural hierarchy. Third, SystemC variables and events cannot be used to connect the ports of different modules. They can only be used either inside the modules or globally. This limits the use of events for scheduling modules and variables for data transfer between modules. SpecC does support scheduling using events and data transfer using variables without any constraints. Fourth, SystemC uses lower-level semantics and syntax to model concepts at higher levels of abstraction. An example is the use of module (which is essentially a structural entity) as the behavioural entity in the specification model. Another example is the use of signals for data transfer. Since the values of signals are updated after a delta cycle delay, using signals to model data transfers may cause timing problems.

Table 3.6 Handel-C data types

Conventional C and Handel-C	Conventional C only	Handel-C only
int	double	chan
unsigned	float	ram
char	union	rom
long		wom
short		mpram
enum		signal
register		chanin
static		chanout
extern		undefined
struct		interface
volatile		<>
void		inline
const		typeof
auto		
signed		
typedef		

Fifth, SystemC is based on the C++ language which is tedious to profile because of the C++ libraries. There are no such limitations with SpecC. Sixth, in the case of SystemC, when static sensitivity is used for scheduling, it affects both the data transfer and the execution sequence scheduling. Therefore, designers can only use dynamic sensitivity for scheduling in the specification model. This is not the case with SpecC.

Handel-C

Handel-C [24] is a programming language that is designed to enable the compilation of programs into synchronous hardware. It is a high-level programming language which targets low-level hardware, most commonly used in the programming of FPGAs. The Handel-C design suite provides support for Altera, Xilinx and Actel FPGAs. Unlike many other design languages that target a specific architecture, Handel-C can be compiled to a number of design languages and then synthesised to the corresponding hardware. The Handel-C compiler can provide output in different formats such as the netlist, VHDL codes and debugging mode that can be used with SystemC or ANSI C for verification. The Handel-C syntax is based on that of conventional C with non-standard extensions to control hardware instantiation with an emphasis on parallelism.

Programmers who are familiar with conventional C would be able to recognise almost all the constructs in the Handel-C language. Handel-C supports all of the primitive integral types provided by ANSI C apart from float, double and long double, signed and unsigned as shown in Table 3.6. It has a range of additional types for creating channels and interfaces between hardware blocks and for specifying memories and signals. In Handel-C, variables are implemented as registers, and the depth of an array must be specified at compile time. Table 3.7 shows the operators

Table 3.7 Handel-C operators

Conventional C and Handel-C	Conventional C only	Handel-C only
* (pointer indirection)	sizeof	select (…)
& (address of)		width (…)
–		@
+		\\
* (multiplication)		<–
/		[:]
%		let … in
<<		
>>		
>		
<		
>=		
<=		
==		
!=		
& (bitwise and)		
^		
|		
?		
[]		
!		
&&		
~		
||		
–>		

supported by Handel-C. Besides the operators supported in ANSI C, other operators that are supported by Handel-C include bit manipulation, range selection and concatenation:

- *Bit manipulation:* Handel-C allows bit manipulation where the number of bits can be take or drop. These are very cheap in hardware since these operators are implemented as wires.
- *Range selection:* If a user wishes to select just a few bits in a register Y, this expression X = Y[n:m] will take bits n to m and store into register X.
- *Concatenation:* Handel-C also allows concatenation of expressions together like expression 1 @ expression 2.

There are no floating-point types (float, double or long double) in Handel-C as floating-point arithmetic is more complex than integer- or fixed-point arithmetic and require more hardware. Handel-C data types are not limited to specific widths. It is the responsibility of the developer to specify the minimum width required to minimise hardware usage. For an example, for the statement "int 5 x", the variable x can hold a value between 1 and 32. There is no automatic conversion between

Fig. 3.18 Sequential and
parallelism in Handel-C

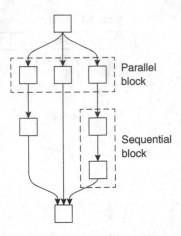

signed and unsigned values in Handel-C; the programmer would have to explicitly
cast them such as:

```
int 12 x;
unsigned int 12 y;
y = (unsigned) x;
```

Since all logic must be expanded at compilation time to generate the hardware,
functions in Handel-C may not be called recursively. Functions can only be called
in expression statements and may not contain any other calls or assignment. The
main() function take no arguments and return no values. However, more than one
main() function is allowed in Handel-C as long as each of them is associated with
a clock. Each loop statement in Handel-C should take at least one clock cycle. For
example, for the following ANSI C codes, the while loop would not be executed if
i was equal to 0.

```
                                              --i;
                                              if (i!=0)
                                                 while (i!=0)
                                              {
                                                 MyFunction (i);
                                                 --i;
   while ((--i)!=0)                            }
   {                                           else
      MyFunction (i);                             delay;
   }
```

Listing 3.1 ANSI C code   **Listing 3.2** Equivalent Handel-C

In Handel-C, assignments and the delay statement take one cycle. All other
operations are on the fly. This allows programmers to manually schedule tasks and
create effective pipelines. By arranging loops in parallel with the correct delays,
a pipeline can massively speed up a program. Since the logic circuit operation is
highly parallel by nature, it is necessary for a design tool to support parallelism as
shown in Fig. 3.18. This can be accomplished in Handel-C by using a par statement,
as opposed to a seq statement where the code is executed sequentially.

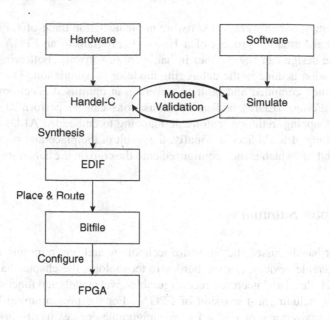

Fig. 3.19 C-to-FPGA design flow

By declaring a variable as a signal, the value of the signal lasts only for the duration of the current clock cycle. Listing 3.3 and 3.4 shows an example of signal usage in Handel-C.

```
signal  unsigned  8 a ;
signal  unsigned  8 b ;

par
{
   a = 7 ;
   b = a ;
}

Results :  a = 0 , b = 7 ;
```

```
signal  unsigned  8 a ;
signal  unsigned  8 b ;

seq
{
   a = 7 ;
   b = a ;
}

Results :  a = 0 , b = 0 ;
```

Listing 3.3 Handel-C signal using "par"   **Listing 3.4** Handel-C signal using "seq"

3.3.5.3 Celoxica DK Design Methodology

The Celoxica DK Design Suite [23] using the Handel-C HDL offers a software flow for algorithm development, optimisation and acceleration in embedded systems. Handel-C allows the designer to verify the designed systems in cycle-accurate simulations. In addition, the DK Design Suites API libraries allow software designers to connect processors and board peripherals to the FPGA logic enabling rapid exploration of hardware and software implementation trade-offs. Figure 3.19 shows the transformation process of a Handel-C program to an FPGA hardware solution. The design can begin either in hardware or software. Following this, the design validation is done in the debugging mode or by simulation. The error-free program is then compiled and built to generate an optimised Electronic Design Interchange Format (EDIF) netlists. Optimisations can be performed including technology mapping, retiming, automatic mapping to embedded ALUs and automatic pipelining of RAM access. Finally, the result of the place-and-route process generates a bit file which is then configured onto the chip on the hardware platform.

3.4 Chapter Summary

This chapter has discussed the hardware technology and programming languages for reconfigurable devices. For the hardware technology, the chapter has covered various system-level architectures, reconfigurable-level models and functional block architectures including a discussion of FPGAs. For the programming languages technology, an overview of HDLs for reconfigurable devices has been presented including lower-level HDL languages like VHDL and Verilog and higher-level HDL C-based languages like SystemC, SpecC and Handel-C. The next chapter will discuss hardware platforms for the WMSNs and the design of a specific processor for the WMSN using the Handel-C HDL.

Chapter 4
FPGA Wireless Multimedia Sensor Node Hardware Platforms

Abstract This chapter presents the designs and implementations for the FPGA wireless multimedia sensor node (WMSN) hardware platforms. Two platforms will be described: a low-cost platform using the Celoxica RC10 FPGA board and a medium-cost platform using the Celoxica RC203E FPGA board. A strip-based low-memory processor based on a modified MIPS architecture will be implemented on the FPGA. For efficient processing, the strip-based MIPS processor contains customised instructions to perform the discrete wavelet transform (DWT). The chapter begins with a discussion of FPGA-based soft-core processors in wireless sensor systems and then moves on to describe the WMSN hardware platforms using the Celoxica FPGA boards. Next, the datapath and control architectures for the strip-based MIPS are discussed. The chapter concludes with an illustrative implementation of the DWT on the hardware platform using the Handel-C hardware description language. The DWT implementation will also be used in the later chapters on event detection and event compression.

4.1 Introduction

While many wireless sensor systems only need to perform a specific task after deployment, situations may arise leading to the need to perform multiple tasks or to change the goals of the system. Often, the processor implemented on the sensor node is non-configurable and may require a full network redeployment. This is especially the case for application-specific integrated circuit (ASIC) sensors. To address this disadvantage, three reconfiguration techniques specifically for FPGA-based soft-core processors to be deployed in the field were proposed in [50]. The first technique is a static method where the reconfiguration software has been preprogrammed into the system. When certain conditions are met, the system will reconfigure itself to suit the intended task. A second method uses a pseudo-static technique to reconfigure the system by means of a wireless medium. Both of these techniques require halting all operations for reconfiguration and the functionalities of the sensor nodes are impaired for a certain time period. The third method, called the dynamic reconfiguration, allows partial system reconfiguration through wireless communication to minimise the impact of hardware changes on the operation of the sensor nodes. In [73], a similar technique termed batch reconfiguration is used to reconfigure the processor of the sensor node on the fly.

The motivations for using FPGA-based soft-core processors in wireless sensor systems can be listed as follows:

- *Computational efficiency.* As compared to serial computing in microcontrollers, parallelism can be easily achieved in FPGAs. Parallel structures in computing architectures lead to computational efficiency as the data throughput is increased.
- *Energy efficiency.* The energy efficiency of a soft-core processor can be viewed as a relationship to its computational efficiency arising from its design structure. The analysis in [31] showed that the power consumption of an FPGA-based system is in direct relation to its architectural design. Also, global clocks in FPGA-based soft-core processors can be varied to obtain the highest performance–power ratio.
- *Reconfigurability.* The FPGAs major resources such as the CLBs and memory blocks are interconnected in a reconfigurable manner to allow architectural changes to be made according to the intended application. By exploiting this feature, soft-core processors of deployed wireless sensor systems can be updated or fully reconfigured remotely through the wireless medium. Reconfiguration techniques allow hardware changes to be made in stages, allowing operation sustainability throughout the reconfiguration period.
- *Supportive role.* In many cases, microcontrollers are unable to provide the computational capacity required for processing complex algorithms. FPGA-based soft-core processors acting as a secondary processor enables these computations to be performed and can be put into the sleep mode to conserve energy when not required. By using a combination of a microcontroller and a FPGA-based soft-core processor configuration, optimised system efficiency can be achieved.

The high power consumption in FPGAs remains a drawback for battery-powered wireless sensor systems. Realising the need for low-powered FPGAs for remote devices, major FPGA manufacturers such as Altera and Xilinx began manufacturing a new range of low-power high-performance FPGAs such as the Altera Stratix-V [7] and the Xilinx Artix-7 [141]. With the availability of low-cost camera sensors and the growing development of wireless multimedia sensor node (WMSNs), resource demanding vision-computing algorithms are likely to be implemented, shifting the processing role in microcontrollers to more computational capable processors. Hybrid microcontroller-FPGA configurations have also been used in more recent visual sensor nodes such as the Micrel Eye platform where simple computations are handled by the microcontroller and the computationally demanding ones by the FPGA. Reconfiguration flexibility, the ease of peripheral integration and the high processing capabilities are evident in current soft-core processors implemented on FPGAs, making them an ideal choice for WMSN hardware platforms.

FPGA-based soft-core processors can be found in the development of wireless sensor systems. In [82], an asynchronous processor is implemented using the FPGA to address the concern of limited operation time in battery-powered wireless sensor nodes. To minimise power consumption, the processor of the sensor node is kept in the sleep mode when no processing is performed. However, in synchronous processors, the global clock remains active even when the sensor node is in the idle state. This causes the battery to be drained regardless of whether any processing is being performed and impacts the operation time significantly. The asynchronous processor uses an event-driven approach to eliminate the need for a global clock and only powers up along with its related modules when an event is detected. An evaluation of the asynchronous processor showed that an AA battery is capable of sustaining its operation for about 60 days. In another FPGA-based soft-core processor implementation for multimedia sensor networks [102], a 16-bit low-speed reduced instruction set computing (RISC) processor is implemented on the Xilinx Spartan-3 XCS1000 FPGA. The processor has a four-stage pipeline architecture and is optimised to perform on-board discrete wavelet transform (DWT) operations prior to image transmission. Hardware comparisons showed that a global clock running at 112.15 MHz consumes approximately four times more power than a clock of 10 MHz. This work concluded that low-speed processors are more suitable for implementation in sensor nodes to reduce power consumption. However, a low global clock rate should only be used when minimal visual computing is involved less it results in the system inability to perform in real-time for more complex visual processing algorithms.

A wireless smart sensor for structural health monitoring proposed in [80] uses an FPGA soft-core processor and a microcontroller with shared resources to increase the processing capability for performing the monitoring operations. In this sensor node architecture, the microcontroller is responsible for managing the I/O controls, communications and performing simple computations. The microcontroller is low power and is always in the active state to perform the data acquisition. The FPGA soft-core processor acts as a secondary processor and remains in the sleep mode until there is a need to perform complex mathematical computations such as the discrete cosine transform (DCT) and the fast Fourier transform (FFT) signal analysis required by the monitoring operation. The inclusion of two different processor designs (hard core and soft core) into the smart sensor design allows the node to continuously acquire data with reduced power consumption for constant monitoring without sacrificing the processing capabilities.

4.2 WMSN Hardware Platforms

This section describes the designs and implementations for the WMSN hardware platforms using the Celoxica FPGA development boards [21]. The platforms are constructed using off-the-shelf components and consist of four main components: a CMOS camera as the imager, an FPGA chip to implement the soft-core processors

Fig. 4.1 Overview of the Celoxica RC10 FPGA board

and its related modules, a networking module for wireless transmission and a main board to provide I/O interconnections between shared resources and ports. Two platforms will be described: a low-cost platform using the Celoxica RC10 FPGA board and a medium-cost platform using the Celoxica RC203E FPGA board.

4.2.1   Low-Cost WMSN Platform Using Celoxica RC10

The Celoxica RC10 board [21] is a platform used for designing, developing and evaluating high-performance FPGA-based applications. Figure 4.1 shows an overview of the RC10 FPGA board. RC10 platforms are fitted with either a standard Xilinx Spartan-3 XC3S-series FPGA chip or a low-power version. Here, the low-power Spartan-3 XC3S1500L is used. The Xilinx Spartan-3 XC3S1500L FPGA is a low-power counterpart to the Spartan-3 XC3S1500 model. For the WMSN platform, the FPGA is configured to implement a soft-core processor based on a modified MIPS architecture. It also provides data transfers between the I/O ports and the internal resources and to allow communication to the wireless module. As the RC10 does not come with any external memory, the FPGA is also used to create a memory block which is global to the processors, the imager and the wireless module.

The Omnivision OV9650 1.3 megapixel colour CMOS camera is linked to the RC10 camera connector and serves as the imager for the WMSN platform. The OV9650 can be used to capture colour images with normal or low-lighting conditions. The colour RGB capture consists of a 5-bit red component, a 6-bit green component and a 5-bit blue component. To reduce memory storage on the FPGA, a greyscale input of 128×128 pixels using only the 128×128 6-bit green component at the centre of a 320×240 QVGA image resolution was used. The 6-bit greyscale

Fig. 4.2 Overview of the Celoxica RC203E FPGA board

approximation has a pixel value range between 0 and 63. Other peripherals such as a joy stick, audio ports, a VGA port, seven-segment displays and a RS-232 port are also fitted on the development board. Compared to other Celoxica development boards, the RC10 plays the role as a low-cost platform. On-board external memory and a display screen are not available on the board.

4.2.2 Medium-Cost WMSN Platform Using Celoxica RC203E

Figure 4.2 shows an overview of the RC203E FPGA board [22]. The RC203E development kit includes a Xilinx Virtex II FPGA device, two external memory banks, a SmartMedia flash memory, a video camera, a touch screen and a RS232 port. Two banks of zero-bus turnaround static random access memory (ZBT SRAM) are available in the RC203E platform. Each memory bank provides a 4 MB storage capacity and is capable of operating at up to 100 MHz. A 16 MB SmartMedia card is provided in the RC203E platform. It is a non-volatile memory that is used for storing BIT files. When power is supplied to the RC203E platform, the FPGA device is configured using the BIT file which is stored in the SmartMedia flash memory. The RC203E board is fitted with a Philips SAA7113H video input processor which enables the FPGA device to capture video data. In the hardware implementation, a 330 line charge-coupled device (CCD) camera provided with the RC203E development kit is used to capture the video frames. These video frames are then stored into the ZBT SRAM for further processing. The RC203E board is also fitted with a 6.4 inch touch screen panel. The video frame captured by the CCD camera can be displayed on the touch screen. A RS-232 serial transmission port is also provided in the RC203E platform. The RS-232 port is capable of receiving as well as transmitting data in an unsigned integer single-byte format.

Fig. 4.3 Overview of the Digi Xbee ZigBee mesh development kit

4.2.3   Digi XBee ZB Zigbee Wireless Module

The WMSN was implemented using the Digi XBee ZNet 2.5 ZigBee® mesh development kit [64]. The XBee ZNet 2.5 radio-frequency (RF) module is an industrial, scientific and medical (ISM) band transceiver which has been designed to provide a low-cost and a low-power wireless networking solution. The ZigBee protocol is based on the IEEE 802.15.4 standard running on a 2.4 GHz band with three additional layers for networking, security and application. The ZigBee protocol uses the mesh network architecture which allows automatic network re-routing if any of the node fails for any reason, thus providing network resilience and error tolerance. Figure 4.3 shows the devices provided with the Digi XBee ZNet 2.5 ZigBee mesh development kit. The kit contains a XBee ZNet 2.5 coordinator, four XBee ZNet 2.5 routers/end devices, two USB interface boards and three RS-232 interface boards. The end device can be used to transmit and receive radio-frequency (RF) data but does not have the capability of data routing. This hardware is attached to the RC10 board through a RS-232 connection and acts as a data transmitter–receiver in the hardware design. Data routing is handled by the router. The router allows data packets to be re-routed in the case of node failure and ensures data transmitted by the end device is propagated to the coordinator. The coordinator acts as the base station with the functions of channel selection for data receiving and transmission as well as assisting data routing. In the WMSN hardware implementation, the coordinator is connected to a host workstation using a USB connection. The bit stream generated by the FPGA module is transmitted by the end device and is propagated throughout the wireless network until it reaches the coordinator.

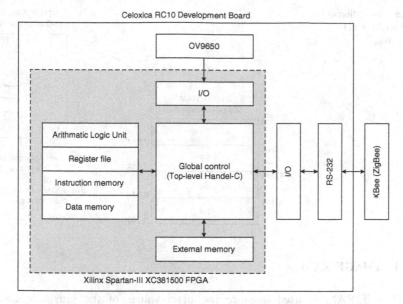

Fig. 4.4 Hardware platform architecture for the RC10 WMSN

4.2.4 Platform Architecture

The hardware platform architecture for the RC10 WMSN is shown in Fig. 4.4. The soft-core processor, the global controls and the external memory are implemented in the FPGA, while peripherals such as the imaging and the networking module are made accessible to the FPGA implementation through the platform abstraction layer (PAL) functions provided with the RC10 and RC203E development boards. The strip-based MIPS processor implements the vision-based algorithms for the WMSN. Hardware implementation of the processors and the global controls are performed using a top-level approach where the behavioural and structural descriptions are specified with Handel-C.

4.2.5 Memory Architecture

The Celoxica RC10 board does not possess any on-board external memory. Thus, all memory requirements have to be fulfilled using the FPGAs own resources and careful design is required to reduce the memory requirements. To meet this requirement, the architecture is implemented with two main memories: IMAGE_RAM and STRIP_RAM. The architecture is designed to operate on a greyscale input of 128×128 pixels but processing is done in smaller strips where an image is first partitioned into strips of 16×128 pixels. For each input image, there will be eight image strips.

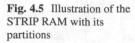

Fig. 4.5 Illustration of the STRIP RAM with its partitions

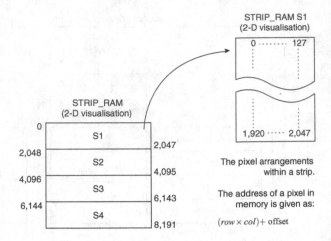

4.2.5.1 IMAGE RAM

The IMAGE_RAM is used to store the pixel values of the greyscale image which is read from the imaging sensor before any processing is performed. The IMAGE_RAM comprises of $128 \times 128 = 16,384$ locations with each location having a width of 16 bits. Basically, the IMAGE_RAM holds the pixel values of the captured image until all processing is completed from the image capture to the transmission. It is reloaded with new pixel values from the next visual capture only after all processing has been completed for the current image. During processing, the IMAGE_RAM is not used except for loading in pixel values of the image strip into the STRIP_RAM.

4.2.5.2 STRIP RAM

The STRIP_RAM is used for all processing operations with the exception of the visual capture. It is used to store the pixel values of the image strip and temporary data during processing. The STRIP_RAM comprises of 8,192 memory locations with each location having a width of 16 bits. In order to perform data transfer and processing operations efficiently, the STRIP_RAM is imagined to be partitioned into four sections. These partitions are referred to as S1, S2, S3 and S4 where S1 denotes partition one and so on. Figure 4.5 illustrates the STRIP_RAM along with its partitions in the 2D visualisation format. The 2D visualisation format will be used to describe information and data values pertaining to the locations in both the IMAGE RAM and the STRIP RAM.

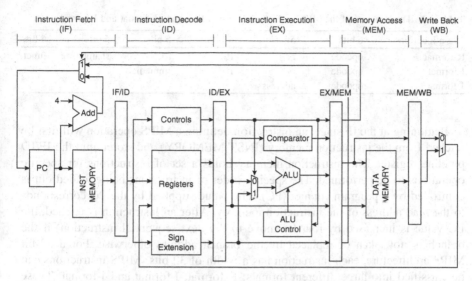

Fig. 4.6 Simplified architecture for a standard five-stage pipelined MIPS processor

4.3 WMSN Processor Architecture

This section describes the strip-based low-memory processor based on a modified MIPS architecture. The processor utilises a four-stage pipeline and a simplified memory architecture. The four-stage pipeline architecture has a simple datapath and also results in a simpler architecture for handling data and control hazards in the pipeline. For efficient processing, the strip-based MIPS processor contains customised instructions to perform the DWT.

4.3.1 Structure of a MIPS Microprocessor and Customisations

The main stages for a standard five-stage pipeline 32-bits MIPS architecture [100] is shown in Fig. 4.6. This MIPS processor which is a form of RISC is used as the base design for the strip-based MIPS processor to be used in the WMSN platforms. As such, it is instructive to briefly consider the design of the standard MIPS processor. The standard MIPS processor consists of five pipeline stages which are designated as instruction fetch (IF), instruction decode (ID), instruction execution (EX), memory access (MEM) and data write back (WB), respectively. The pipelined structure allows for the overlapping of multiple instructions during execution. Pipelining allows the instruction throughput to be increased at the cost of a slight increase in latency.

Table 4.1 The structure of the MIPS instruction code in R-format, I-format and J-format

Bits allocation	6 bits	5 bits	5 bits	5 bits	5 bits	6 bits
R-format	opcode	rs	rt	rd	shamt	funct
I-format	opcode	rs	rt	immediate		
J-format	opcode	address				

Beginning at the IF stage, an instruction defining a MIPS operation will first be fetched from the instruction memory (INST MEMORY) and stored into the IF/ID pipeline register. The instruction memory holds a set of instructions for specific operations to be performed. The specific order of instructions being loaded out is controlled by a program counter (PC). The value supplied by the PC corresponds to the read address of the instruction memory. After an instruction is fetched, the PC value is increased by four (to move to the next sequential instruction) if the branch is not taken or replaced by the branch address otherwise. For a 32-bit MIPS architecture, each instruction has a width of 32 bits. MIPS instructions can be classified into three different formats: R-format, I-format and J-format. These instruction formats are summarised as shown in Table 4.1. From the structure of the formats, it can be seen that an instruction code is divided into several fields which can consist of an opcode, operands (Rs, Rt, Rd), a shift amount (Shamt) or immediate value (Immediate) and a function code (Funct). The operation code (opcode) is used to determine the structure of the instruction.

In the ID stage, the instruction code stored in the IF/ID pipeline register is decoded to provide the necessary signals for controls, operands as well as the type of operation to be performed in the EX stage. In order for the MIPS to generate the correct signals, the format of the instruction has to be determined. This is performed by checking the 6-bit operation code (opcode) of the instruction. The opcode is also used to generate the control signals for the multiplexers in the subsequent stages. If the instruction code is determined to be a R-format (arithmetic and logical operation) type, two items of data (DataRead1 and DataRead2) will be read out from the Registers using the address supplied by the rs and rt fields. These items of data are then stored into the ID/EX pipeline register for processing in the next stage. The type of operation to be performed on these two data is specified by the function code (Funct). The shift amount (Shamt) of the R-format is used to specify the number of shifts for bitwise operations. The Rd field supplies the write address for processed data to be written back into a specific register.

For the I-format instruction, the Rd, Shamt and Funct fields are replaced with an immediate value or a branch address depending on the operation specified by the opcode. The immediate value is used when the operation is a perform immediate arithmetic. For example, the Add Immediate operation can immediately add a value specified by the immediate field to DataRead1 without having to store that specific value into a register. Another important operation of the I-format is the branch operation. This allows the PC to shift to specific instructions specified by the branch address field. The immediate value or the branch address is only 16 bits in width. Therefore, it has to be padded to 32 bits wide through the sign extension before storing into the ID/EX pipeline register. The J-format instruction is used when there is a need to perform an absolute jump in the instruction code. Hence, it has only two fields: the opcode to generate the controls for the multiplexers and an address field to specify which location the PC should point to.

In the EX stage, the data pair DataRead1 and DataRead2 from the Registers or the DataRead1 and the padded immediate value will be used as input into the arithmetic logic unit (ALU) depending on the format type. The secondary input to the ALU is selected by a multiplexer with its select control generated in the ID stage. The controls from the ID stage also supply an ALU operation code which determines whether the ALU should perform a branch computation, an immediate arithmetic or a R-format-type operation based on the function code. At the same time, the comparator performs an equality comparison on DataRead1 and DataRead2 to determine whether a branch should be taken in the case of a branch instruction. After processing if the ALU is completed, the result is stored into the EX/MEM pipeline register. This result can be either used as a read/write address for data reading/writing from/to the Data Memory in the MEM stage or set as a bypass to the MEM/WB pipeline register for data write back to the Registers.

Researchers have proposed modifications to the basic MIPs architecture to improve or customise it for specific applications. In the work by [128], a new branch mechanism that can remove the stalls after a branch instruction is proposed. In the standard MIPS architecture, two clock cycles are needed to determine whether a branch operation should be taken or not. Also, the MIPS architecture is reduced from a five-stage to a four-stage pipeline. This can help to improve the efficiency of code execution and reduce the complexity. Another similar approach is proposed by [58] to improve the branch mechanism as well. The basic idea of these two approaches is to pre-calculate the branch address and store into a specific memory. The work in [147] proposed a MIPS architecture that can reduce the number of stalls. Instead of reading out one instruction code at a time, the proposed architecture can extract three instruction codes simultaneously and detect whether stalls are required. If the condition is true, the sequence of executing the code will be altered to avoid the need of inserting the stalls. In addition, dedicated multiplier and modular arithmetic modules have also been incorporated to improve the processing for cryptography applications. The work proposed by [107] has focused on introducing several new instructions that are custom designed for an application-specific human face detection application.

Recent modifications on the processor have been focusing not only on the architecture itself but also on extending the number of cores. In the work proposed by [10], an approach is adopted to speed up the decoding process for motion JPEG. In this case, the core is implemented using the MIPS architecture and multiple cores are joined together to improve the performance. The work in [41] proposed a dual-core processor with the target of achieving fault tolerance in the processor design. The work in [48] focuses on reducing the power consumption of the MIPS architecture. This is achieved by eliminating all the unwanted transitions. In this case, the data will be passed through to the next stage if the current stage does not perform any processing on the incoming data. Another compact MIPS architecture proposed by [76] that uses very few resources has been realised as a soft-core processor. This architecture is primarily targeted for implementation in resource constrained hardware environments.

4.3.2   Strip-Based MIPS Processor Architecture

This section will describe the design of the strip-based MIPS processor architecture in terms of its instruction set design, datapath design and control design.

4.3.2.1   Instruction Set Design for Strip-Based Processor

The strip-based MIPS processor only uses the R and I instruction formats. The J-format is not used. The instruction set design is divided into two parts: a base instruction set and an extended instruction set. The base instruction set contains a selected subset of the standard MIPS instructions. The extended instruction set contains custom instructions to facilitate the processing for the visual processing applications such as the DWT.

Base Instruction Set

The base instruction set contains a selected subset of the standard MIPS instructions. To reduce the hardware complexity, only 13 instructions are selected from the standard MIPS instructions. The base instruction set is shown in Table 4.2. These instructions allow the processor to perform basic operations such as addressing, arithmetic and memory accesses. In Table 4.2, under the Example column, the notations $R1, $R2 and $R3 represent the register numbers specified by the instruction bits 25:21, 20:16 and 15:11, respectively. In cases where fewer than three registers are required in an operation, the unused field is indicated by the X notation. The unused field is normally set to zero. For architecture simplicity, the shift instructions are implemented only for a single bit shift.

Extended Instruction Set

The extended instruction set contains custom instructions to facilitate the processing for the visual processing applications such as the DWT. There are five custom

Table 4.2 Extended instruction set for strip-based MIPS processor

Operation detail	Code	Format	Example	Remarks
Arithmetic operation				
Addition	ADD	R	$R1, $R2, $R3	$R3 = $R1 + $R2
Subtraction	SUB	R	$R1, $R2, $R3	$R3 = $R1 − $R2
Addition immediate	ADDI	I	$R1, $R2, Val	$R2 = $R1 + Val
Logical operation				
Logical AND	AND	R	$R1, $R2, $R3	$R3 = $R1&$R2
Logical OR	OR	R	$R1, $R2, $R3	$R3 = $R1 | $R2
Logical NOT	NOT	R	$R1, X, $R3	$R3! = $R1
Shift Left Logical	SLL	R	$R1, X, $R3	$R3 = $R1 << 1
Shift Right Logical	SRL	R	$R1, X, $R3	$R3 = $R1 >> 1
Set on Less Than	SLT	R	$R1, $R2, $R3	$R3 = \begin{cases} 1, & \text{if } (\$R2 > \$R1) \\ 0, & \text{otherwise} \end{cases}$
Memory addressing				
Load Word	LW	I	$R1, $R2, X	$R2 = Data Memory [$R1]
Store Word	SW	I	$R1, $R2, X	Data Memory [$R1] = $R2
Memory addressing				
Branch on Equal	BEQ	I	$R1, $R2, $R3	PC = $R3, if $R1 = $R2
Branch on Not Equal	BNE	I	$R1, $R2, $R3	PC = $R3, if $R1 ≠ $R2

Table 4.3 Extended instruction set for strip-based MIPS processor

Operation detail	Code	Format	Example	Remarks
Centre-surround operation				
Absolute	ABS	R	$R1, X, $R3	$R3 =| $R1 |
Absolute on subtraction result	ABSS	R	$R1, $R2, $R3	$R3 =| $R1 − $R2 |
DWT operation				
Shift right on addition result	ADDS	R	$R1, $R2, $R3	$R3 = ($R1 + $R2) >> 1
Shift right twice on addition result	ADDSS	R	$R1, $R2, $R3	$R3 = ($R1 + $R2+2) >> 2
DWT coefficient address rearrange	DWTA	R	$R1, $R2, $R3	$R3 = New Address ($R1, $R2)

instructions in the extended instruction set as shown in Table 4.3. These custom instructions do not operate on immediate values or are involved with the memory access and are categorised under the R-format instructions. The listed customised instructions allow multiple operations to be performed in a single clock cycle and have been selected for visual processing operations. This not only will reduce computation time but will also simplify software-related tasks. An example of this is the DWTA instruction which facilitates the addressing calculations for performing the DWT. In the DWTA instruction, $R2 defines the current decomposition level, $l = 1, 2, 3, 4$. The use of this instruction will be further discussed in Sect. 4.4.

Table 4.4 Machine code for R-format instructions

Instruction	Opcode	Rs	Rt	Rd	Shamt	Funct
ADD	000000	0–31	0–31	0–31	0	100000
SUB	000000	0–31	0–31	0–31	0	100010
AND	000000	0–31	0–31	0–31	0	100100
OR	000000	0–31	0–31	0–31	0	100001
NOT	000000	0–31	0	0–31	0	100110
SLL	000000	0–31	0	0–31	0	000000
SRL	000000	0–31	0	0–31	0	000010
SLT	000000	0–31	0–31	0–31	0	101010
ABS	000000	0–31	0	0–31	0	000100
ABSS	000000	0–31	0–31	0–31	0	000101
ADDS	000000	0–31	0–31	0–31	0	110000
ADDSS	000000	0–31	0–31	0–31	0	110010
DWTA	000000	0–31	0–31	0–31	0	111000

Table 4.5 Machine code for I-format instructions

Instruction	Code	Rs	Rt	Immediate
LW	100011	0–31	0–31	0
SW	101011	0–31	0–31	0
BEQ	000100	0–31	0–31	$0\text{-}(2^{16}-1)$
BNE	000101	0–31	0–31	$0\text{-}(2^{16}-1)$
ADDI	001000	0–31	0–31	$0\text{-}(2^{16}-1)$

Machine Instructions

To communicate with the processor, the instructions have to be given a specific machine code. Tables 4.4 and 4.5 show the machine code assignments for the 13 R-format instructions and the 5 I-format instructions, respectively. The register operands (Rs, Rt and Rd) can take a value from 0 to 31 to address each of the 32 registers.

4.3.2.2 Datapath Design for Strip-Based Processor

The datapath design for the strip-based MIPS processor is shown in Fig. 4.7. The design is a four-stage pipeline processor architecture and contains the following stages: IF, ID, instruction execution and memory access (EXMEM) and data write back (WB). Each stage will be discussed in the next few sections.

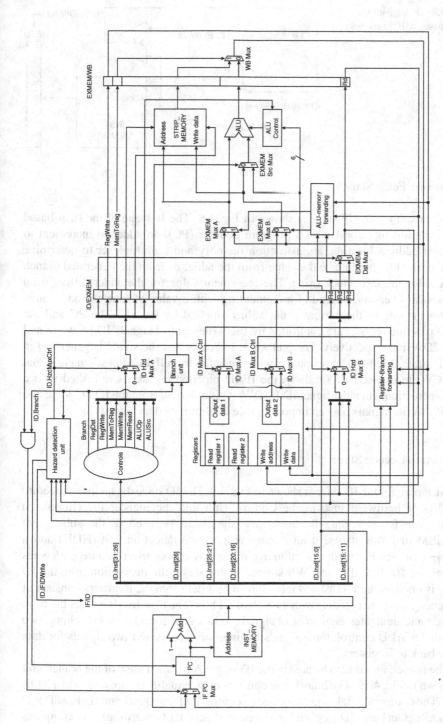

Fig. 4.7 Datapath design for strip-based MIPS processor

Fig. 4.8 IF datapath for
strip-based MIPS processor

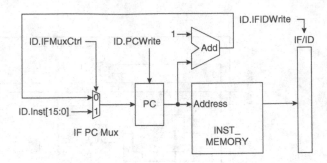

Instruction Fetch Stage

The datapath for the IF stage is shown in Fig. 4.8. The IF stage of the strip-based
MIPS architecture consists of a program counter (PC), an adder to increment to
the next address location, an instruction memory and a multiplexer to determine
whether the PC should take the value from the adder or from the generated branch
address for the next clock cycle. The increment value for the adder differs from
the standard version. Instead of incrementing with a value of four, the adder now
increments one to the current count value supplied by the PC. The PC and the
IF/ID pipeline register are controlled by the write control signals ID.PCWrite and
ID.IFIDWrite, respectively. The prefix ID. indicates that the signal is generated in
the ID stage. The PC is only updated with the value from IF.PCMux when the signal
ID.PCWrite is asserted. Similarly, the IF/ID pipeline register is refreshed with a
new instruction when the signal ID.IFIDWrite is asserted. Both the ID.PCWrite and
ID.IFIDWrite signals play an important role to eliminate data hazards.

Instruction Decode Stage

The datapath for the ID stage is shown in Fig. 4.9. The ID stage datapath architecture
consists of hardware such as the Control Unit and the register file (Registers)
responsible for providing the necessary signals to be used in the subsequent
EXMEM and WB stages. It also contains a hazard detection unit (HDU) and a
Register-Branch Forwarding Unit to avoid hazards arising from data dependencies
among the ID, EXMEM and WB stages. In this stage, the instruction from the ID
stage is decoded and translated into control and data signals. Registers consists of
32 general-purpose registers with a width of 32 bits. The Control Unit takes the 6-bit
opcode and generates eight control signals: one signal for address branching, two
signals for ALU control, three signals for memory access and two signals for data
write back to Registers.

The branch unit is also located in the ID stage. A detailed view of the branch unit
is shown in Fig. 4.10. The branch unit consists of an equality comparator and a XOR
gate. The comparator takes in two values as input, ID.ReadData1 and ID.ReadData2
(or ID.RegOut1 and ID.RegOut2 in the case of no data forwarding), and compares

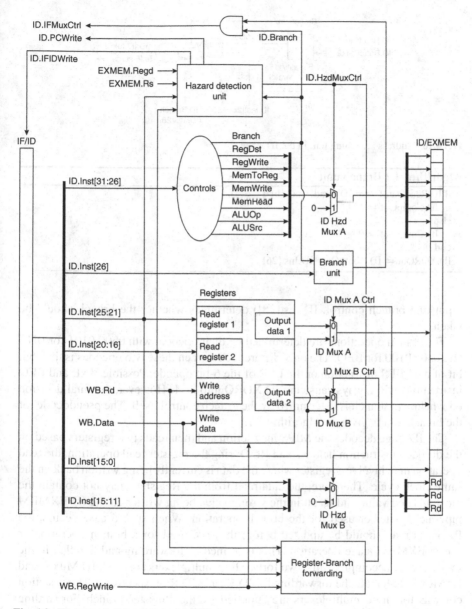

Fig. 4.9 ID datapath for strip-based MIPS processor

them for equality. If both the values are equal, the comparator output ID.CompOut takes a value of 1, else it takes a value of 0. The XOR gate then performs an exclusive OR operation on the ID.CompOut value with the 26th bit of the MIPS instruction (ID.Inst[26]). The output of the XOR is input into a logic AND in which another

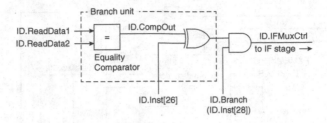

Fig. 4.10 Address branching unit in the ID stage

Algorithm 4.1 Branch unit

---
if ID.ReadData1 == ID.ReadData2 **then**
    ID.CompOut = 1
else
    ID.CompOut = 0
end if
ID.XOROut = ID.CompOut ^ ID.Inst[26]

---

input (the branch control, ID.Inst[28]) determines whether the branch should be taken.

The branch operations are determined by their opcode, with 0b00 0101 for BNE and 0b00 0100 for BEQ. The only difference between these two opcodes is the 26th bit of the MIPS instruction or the LSB of the 6-bit opcode. Possible BNE and BEQ instructions will always cause the ID.XOROut to be 1. However, the final decision on whether to branch is determined by the branch control itself. The pseudocode for the branch unit is given in Algorithm 4.1.

The ID stage decodes the MIPS instruction and then reads two registers based on the decoded instruction field Rs and Rt. During the register read operation, the read operation may involve a register where the data is currently being written back on the same clock cycle. Therefore, the data read from the Registers may not contain the most updated value and result in the wrong value being written to the ID/EXMEM pipeline register or used for the branch operation. When such a case occurs, the Registers read should be updated before they are used for a branch operation or for an EXMEM stage operation. This is termed as forwarding and handled in the processor. To accomplish this, two forwarding multiplexers are used, ID Mux A and ID Mux B, to handle the forwarding for ID.Rs and ID.Rt, respectively. The selection controls for these multiplexers are supplied by the Register-Branch Forwarding Unit. This unit compares the ID register source fields, ID.Rs and ID.Rt, with the register destination WB.Rd in the write back stage. If either ID.Rs or ID.Rt matches the register specified by WB.Rd and the WB.RegWrite is asserted, then the forwarding unit asserts the appropriate multiplexer control signals to select the WB.Data which will replace the register read data. The conditions for the multiplexer controls can be found in Algorithms 4.2 and 4.3. In the case where branching does not occur, the forwarding multiplexers forward the most updated values to the ID/EXMEM pipeline register.

Algorithm 4.2 Register-branch forwarding unit for ID Mux A

---
if (WB.RegWrite == 1) && (WB.Rd != 0) && (ID.Inst[25:21] == WB.Rd) **then**
    ID.MuxACtrl = 1
else
    ID.MuxACtrl = 0
end if

Algorithm 4.3 Register-branch forwarding unit for ID Mux B

---
if (WB.RegWrite == 1) && (WB.Rd != 0) && (ID.Inst[20:16] == WB.Rd) **then**
    ID.MuxBCtrl = 1
else
    ID.MuxBCtrl = 0
end if

Algorithm 4.4 Hazard detection unit

---
if (ID.Branch == 1)
&& (((ID.Rs == EXMEM.Rs) && (ID.Rs != 0) && (ID.Rs != 1))
|| ((ID.Rs == EXMEM.Regd) && (ID.Rs != 0) && (ID.Rs != 1))
|| ((ID.Rt == EXMEM.Rs) && (ID.Rt != 0) && (ID.Rt != 1))
|| ((ID.Rt == EXMEM.Regd) && (ID.Rt != 0) && (ID.Rt != 1))) **then**
    ID.PCWrite   = 0
    ID.IFIDWrite = 0
    ID.HzdMuxCtrl = 1
else
    ID.PCWrite   = 1
    ID.IFIDWrite = 1
    ID.HzdMuxCtrl = 0
end if

The forwarding hardware eliminates data hazards for ALU operations and memory accesses. However, data hazards may occur during a branch operation which uses a register that is being operated upon in the EXMEM stage in the same clock cycle. A HDU is included in the processor design to detect branch operations that have dependency on EXMEM data. The HDU takes in as input four signals: ID.Inst[25:21] (ID.Rs), ID.Inst[20:16] (ID.Rt), EXMEM.Rs and EXMEM.Regd. It compares the IDsignals with the EXMEM signals for dependencies. If a data dependency is found, the HDU disasserts the ID.PCWrite and ID.IFIDWrite signals. This stops all PC and IF/ID pipeline register writes for one cycle so that in the next cycle (after stalling), the PC value and the branch instruction remain. At the same time, ID.HzdMuxCtrl is asserted to select a value of zero for the control, ID.Rs, ID.Rt and ID.Rd signals to be written into the ID/EXMEM pipeline register. On the next clock cycle, the branch operation will resume with the correct values forwarded from the WB stage. The EXMEM stage effectively performs a no operation (NOP). Algorithm 4.4 shows the condition for detecting a data hazard in a branch operation.

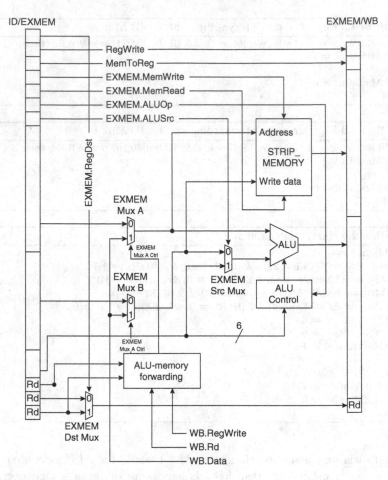

Fig. 4.11 EXMEM datapath for strip-based MIPS processor

Instruction Execution and Memory Access Stage

The hardware architecture for the EXMEM stage is very similar to the EX and MEM stages in the standard MIPS with several simplifications to the controls due to the combination of these two stages. The datapath for the EXMEM stage is shown in Fig. 4.11. In the standard MIPS, memory operations require an offset addition in the EX stage with the ALU output to be used as the address to access the memory. Memory access in the MEM stage is bypassed when the memory controls are not asserted. In the strip-based MIPS, the memory is accessed only when the memory control is asserted. An ALU operation is performed regardless of whether the instruction received is an ALU operation or a memory access operation. The correct data to be written back to the Registers is selected by the MemToReg control.

Algorithm 4.5 ALU-memory forwarding unit for EXMEM Mux A

if (WB.RegWrite == 1) && (WB.Rd != 0) && (EXMEM.Rs == WB.Rd) **then**
    EXMEM.MuxACtrl = 1
else
    EXMEM.MuxACtrl = 0
end if

Algorithm 4.6 ALU-memory forwarding unit for EXMEM Mux B

if (WB.RegWrite == 1) && (WB.Rd != 0) && (EXMEM.Rt == WB.Rd) **then**
    EXMEM.MuxBCtrl = 1
else
    EXMEM.MuxBCtrl = 0
end if

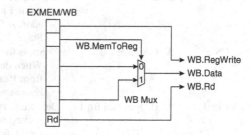

Fig. 4.12 WB datapath for strip-based MIPS processor

For data forwarding in this stage, the ALU-Memory Forwarding Unit is used. The conditions for the multiplexer controls are described in Algorithms 4.5 and 4.6.

Instruction Write Back Stage

The datapath for the data write back (WB) stage is shown in Fig. 4.12. In this stage, data written back to the Registers is performed depending on whether the control signal WB.RegWrite is asserted or de-asserted. The data to be written back comes from either the ALU or the memory in the EXMEM stage and is selected with the control signal WB.MemToReg.

4.3.2.3 Control Design for Strip-Based Processor

The instruction operation codes (opcodes) and control signals for the strip-based MIPS processor are shown in Table 4.6. Columns I31 to I26 show the opcodes for the operations with I31 being the most significant bit of the 32-bit MIPS instruction. The opcode for the I-format instructions is individually listed since the type of operation is determined by the opcode only. For the R-format instructions, the operation to be performed is determined by the function code rather than the opcode. All R-format instructions use a 6-bit binary value of zeros. There are eight control

Table 4.6 List of instruction opcodes and their control signals

Instruction	I_{31}	I_{30}	I_{29}	I_{28}	I_{27}	I_{26}	Branch	ALU Src	ALU Op1	ALU Op0	Mem Read	Mem Write	Mem ToReg	Reg Dst
R-format	0	0	0	0	0	0	0	0	1	0	0	0	0	1
LW	1	0	0	0	1	1	0	1	0	0	1	0	1	1
SW	1	0	1	0	1	1	0	1	0	0	1	0	X	X
BNE	0	0	0	1	0	1	1	0	0	1	0	0	X	0
BEQ	0	0	0	1	0	0	1	0	0	1	0	0	X	0
ADDI	0	0	1	0	0	0	0	1	0	0	0	0	0	1

Table 4.7 Description and function of major control signals

Signal	Description	Function
Branch	Branch	When asserted, causes the IF multiplexer to select from Input 1 to update the PC with a specific branch address from the Sign Extended signal in the ID Stage
ALUSrc	ALU Source	Controls the multiplexer output into the ALU Input 2. When de-asserted, input to the ALU Input 2 comes from ReadData2, else from the immediate signal; both supplied by the ID/EXMEM pipeline register
ALUOp1	ALU Operation Bit 1	Determines whether the ALU should perform an immediate addition (0b00) and/or operation based on the function code supplied (0b10)
ALUOp0	ALU Operation Bit 0	Determines whether the ALU should perform an immediate addition (0b00) and/or operation based on the function code supplied (0b10)
MemRead	Memory Read	When asserted, allows the Data Memory to be read from
MemWrite	Memory Write	When asserted, allows the Data Memory to be written to
MemtoReg	Memory tp Register	When asserted, allows read data from Data Memory to be written back to the registers. When de-asserted, the ALU output is to be written back
RegDst	Register Destination	When asserted, allows the Registers to be written to

signals in the strip-based MIPS architecture: Branch, ALUSrc, ALUOp1, ALUOp0, MemRead, MemWrite, MemToReg and RegDst. The descriptions of the control signals are shown in Table 4.7.

4.4 Discrete Wavelet Transform and Handel-C Implementation

The DWT is the mathematical core for many visual information processing schemes. Traditional two-dimensional (2D) DWT first performs row filtering on an image followed by column filtering. This gives rise to four wavelet sub-band decompositions: the low–low (LL) sub-band containing the low-frequency content of an image in both the horizontal and vertical dimensions, the high–low (HL) sub-band containing the high-frequency content of an image in the horizontal dimension and low-frequency content of an image in the vertical dimension, the low-high (LH) sub-band containing the low-frequency content of an image in the horizontal dimension and high-frequency content of an image in the vertical dimension and the high–high (HH) sub-band containing the high-frequency content of an image in both the horizontal and vertical dimensions. Each of the wavelet coefficients in the LL, HL, LH and HH sub-bands represents a spatial area corresponding to approximately a 2×2 area of the original image. For a N-scale DWT decomposition, the coarsest LL sub-band is further decomposed. As a result, each coefficient at a coarser scale represents a larger spatial area of the image but at a narrower band of frequency. Two approaches that can be used to implement the DWT are the convolution-based filtering method and the lifting-based filtering method. The lifting-based DWT is preferred over the convolution-based DWT for hardware implementation for several reasons:

- *It gives a better computational efficiency.* It was reported in [38] that the lifting-based DWT can save up to 50 % of the computational requirement compared to the convolution-based approach.
- *It requires less memory buffer.* The in-place computing feature of lifting requires no extra memory buffer.
- *It provides an integer-to-integer transformation.* The property of integer-to-integer lifting transformation enables lossless visual processing.
- *It requires a less complex hardware implementation.* Only simple logic such as addition, subtraction and shifter are needed. It also requires a less complicated inverse wavelet transform.

4.4.1 Lifting-Based DWT

In visual processing, a 2D-DWT is performed to decompose the image into multi-resolution wavelet sub-bands of different frequency content. For hardware implementation, the lifting-based DWT using the reversible Le Gall 5/3 filter [11] was adopted. This filter is selected since it provides for a lossless transformation and requires a less complex hardware implementation. Figure 4.13 shows the hardware architecture for the lifting-based 5/3 DWT.

In the implementation of the 5/3 DWT, three computation operations (addition, subtraction and shift) are needed. The lifting process is built based on the split, prediction and updating steps. The input sequence $X[n]$ is first split into odd (high-frequency) and even (low-frequency) components for the horizontal filtering process. In the prediction phase, a high-pass filtering is applied to the input signal which results in the generation of the detailed coefficient $H_{\text{spatial}}[2n + 1]$. In the updating phase, a low-pass filtering is applied to the input signal which leads to the generation of the approximation coefficient $L_{\text{spatial}}[2n]$. Likewise, for the vertical

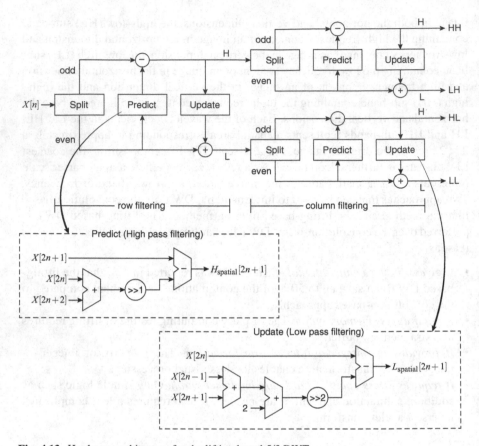

Fig. 4.13 Hardware architecture for the lifting-based 5/3 DWT

filtering process, the split, prediction and updating steps are repeated for both the $H_{\text{spatial}}[2n+1]$ and $L_{\text{spatial}}[2n]$ coefficients. Equations (4.1) and (4.2) give the lifting implementation for the 5/3 filter used in JPEG 2000.

$$H_{\text{spatial}}[2n + 1] = X[2n + 1] - \left[\frac{X[2n] + X[2n + 2]}{2} \right] \tag{4.1}$$

$$L_{\text{spatial}}[2n + 1] = X[2n] - \left[\frac{H[2n - 1] + H[2n + 2] + 2}{4} \right] \tag{4.2}$$

Using a step-by-step computation, the high-pass and low-pass outputs can be computed in a standard MIPS processor as shown in Algorithms 4.7 and 4.8, respectively. The high-pass and the low-pass output computations would take six and eight clock cycles each, respectively.

Algorithm 4.7 Step-by-step high-pass output computation in MIPS

1: Load $X[2n]$ from STRIP_RAM to register $R18
2: Load $X[2n + 1]$ from STRIP_RAM to register $R19
3: Load $X[2n + 2]$ from STRIP_RAM to register $R20
4: Add the values of $R18 and $R20 and store into a temporary register $R21
5: Bitshift the value of $R21 to the right by 1 (equivalent to divide by 2)
6: Subtract the value in $R21 from $R19 and store the result back into $R19

Algorithm 4.8 Step-by-step low-pass output computation in MIPS

1: Load $X[2n − 1]$ from STRIP_RAM to register $R17
2: Load $X[2n]$ from STRIP_RAM to register $R18
3: Load $X[2n + 1]$ from STRIP_RAM to register $R19
4: Add the values of $R17 and $R19 and store into a temporary register $R21
5: Add 2 to $R21 and store result into $R21
6: Bitshift the value of $R21 to the right by 1 (equivalent to divide by 2)
7: Bitshift the value of $R21 to the right by 1 (equivalent to divide by 2)
8: Subtract the value in $R21 from $R18 and store the result back into $R19

Algorithm 4.9 Step-by-step high-pass output computation in MIPS using ADDS

1: Load $X[2n]$ from STRIP_RAM to register $R18
2: Load $X[2n + 1]$ from STRIP_RAM to register $R19
3: Load $X[2n + 2]$ from STRIP_RAM to register $R20
4: Add the values of $R18 and $R20, then bitshift the result to the right by 1 and store into $R21
   (ADDS instruction)
5: Subtract the value in $R21 from $R19 and store the result back into $R19

Algorithm 4.10 Step-by-step low-pass output computation in MIPS using ADDSS

1: Load $X[2n − 1]$ from STRIP_RAM to register $R17
2: Load $X[2n]$ from STRIP_RAM to register $R18
3: Load $X[2n + 1]$ from STRIP_RAM to register $R19
4: Add the values of $R17, $R19, and 2, then bitshift the result to the right by 2 and store into
   $R21 (ADDSS instruction)
5: Subtract the value in $R21 from $R18 and store the result back into $R19

Since the filtering process involving these two computations is applied to every pixel over four decomposition levels, the strip-based processor uses two algorithm-specific MIPS instructions ADDS and ADDSS to reduce the clock cycles required for the high-pass and low-pass output computations. Algorithms 4.9 and 4.10 show the computation changes made with the ADDS and ADDSS instructions, respectively. With the ADDS and ADDSS instructions, both high- and low-pass outputs can be computed using five clock cycles leading to a 16.7 % and 37.5 % reduction in clock cycles for each pixel filtered. In this case, $R17 to $R21 are the five registers used to store the computation result.

Algorithm 4.11 Stage 0: data copy

Define:
$R3 - read address pointer (reads from S1)
$R4 - write address pointer (writes to S3)
$R5 - number of pixels to be copied
$R6 - pixel counter
Initialisations: $R3 = 0; $R4 = 4096; $R5 = 2048; $R6 = 0;
while $R6 < $R5 **do**
    $R7 = STRIP_RAM[$R3]
    $R3 = $R3++
    STRIP_RAM[$R4] = $R7
    $R4 = $R4++
end while

4.4.2   Hardware Architecture for DWT

This section describes the hardware implementation of the 5/3 DWT on the WMSN hardware platform. The implementation consists of two stages: data copy and DWT. In the implementation, these stages are translated and implemented as a set of MIPS binary instructions on the strip-based processor.

4.4.3   Stage 0: Data Copy

In Stage 0, the top-level control copies an image strip from the IMAGE_RAM to STRIP_RAM. At Stage 0 and before any processing is carried out, the image strip to be processed is located in STRIP RAM S1. Since the data in the S1 partition should not be modified as it may be used at a later stage, the image strip has to be copied out to another working partition. In Stage 1 (DWT), it is preferable for the wavelet coefficients to be in partition S2 so that the partitions can be used in an orderly rotating manner in the subsequent stages. Therefore, the pixel values of the image strip in S1 are copied to partition S3 before the Stage 1 operation is performed. Algorithm 4.11 shows the data copy for Stage 0.

4.4.4   Stage 1: Discrete Wavelet Transform

Stage 1 performs a four-level lifting-based 5/3 DWT decomposition. Algorithm 4.12 shows the DWT for Stage 1. Prior to Stage 1, Stage 0 copies the image strip pixel values to be decomposed to partition S3. Then in Stage 1, the row filtering is first performed on the image strip rows, operating on two or three pixel values at a time depending on if the pixel is located on the image borders.

Algorithm 4.12 Stage 1: discrete wavelet transform

1: **Define:**
2: $R3 - strip image row size; $R4 - strip image column size
3: $R5 - number of pixels in strip image; $R6 - DWT level l
4: $R7 - read address pointer (reads from S3)
5: $R8 - write address pointer (writes to S3)
6: $R9 - row counter; $R10 - column counter
7: $R12 - offset to shift address back to S1 (for DWT address rearrangement)
8: $R17 - DWT filter 1 $[2n - 1]$; $R18 - DWT filter 2 $[2n]$
9: $R19 - DWT filter 3 $[2n + 1]$; $R20 - DWT filter 4 $[2n + 2]$
10: $R26 - offset to shift address back to S2 (for Stage 1 output writing)
11: **Initialisations:** $R3 = 16$; $R4 = 128$; $R5 = 2048$; $R6 = 1$
12: **Initialisations for row filtering:** $R7 = 4096$; $R8 = 4096$; $R9 = 0$; $R10 = 0$
13: **while** $R9 < $R3 **do**
14:     **while** $R10 < $R4 **do**
15:         **STEP 1: Check for filtering at right image border**
16:         **do** right image border check (Algorithm 4.13)
17:         **STEP 2: Row Filtering**
18:         **do** row filtering (Algorithm 4.14)
19:             $R10 = $R10 + 2$
20:     **end while**
21:     $R9 \ = $R9++$
22:     $R10 = 0$
23: **end while**
24: **Initialisations for col filtering:** $R7 = 4096$; $R8 = 4096$; $R9 = 0$; $R10 = 0$; $R12 = 4096$; $R17 = 0$; $R18 = 0$; $R19 = 0$; $R20 = 0$; $R26 = 2048$
25: **while** $R10 < $R4 **do**
26:     **while** $R9 < $R3 **do**
27:         **STEP 3: Check for filtering at bottom image border**
28:         **do** bottom image border check (Algorithm 4.15)
29:         **STEP 4: Column Filtering**
30:         **do** column filtering (Algorithm 4.16)
31:             $R9 \ = $R9 + 2$
32:     **end while**
33:     $R9 \ = 0$
34:     $R10 = $R10 + 2$
35:     $R7 \ = $R10$
36:     $R7 \ = $R7 + $R12$
37:     $R8 \ = $R7$
38: **end while**
39: **STEP 5: Update for next DWT level**
40: **do** update data and parameters (Algorithm 4.17)

4.4.4.1 Computation of High-Pass and Low-Pass Outputs

For each pixel at location n read from S3, a border check is performed to check whether the next pixel $[n + 1]$ lies on the right image border. If the next pixel is found lying on the right image border, then only two pixels are loaded into the DWT filter registers $R18 and $R19, respectively, else three pixels will be loaded

---

Algorithm 4.13 Stage 1: DWT (row-filtering border check)

1: **if** filtering at right border **then**
2:      $R18 = STRIP\_RAM[$R7]
3:      $R7 = $R7++
4:      $R19 = STRIP\_RAM[$R7]
5:      $R7 = $R7++
6: **else**
7:      $R18 = STRIP\_RAM[$R7]
8:      $R7 = $R7++
9:      $R19 = STRIP\_RAM[$R7]
10:     $R7 = $R7++
11:     $R20 = STRIP\_RAM[$R7]
12: **end if**

Algorithm 4.14 Stage 1: DWT (row filtering)

1: **if** $R10 == 0 **then**
2:      $R21 = ($R18 + $R20) >> 1
3:      $R19 = $R19 − $R21
4:      $R21 = ($R19 + $R19 + 2) >> 2
5:      $R18 = $R18 + $R21
6:      $R17 = $R19
7:      STRIP_RAM[$R8] = $R18
8:      $R8 = $R8++
9:      STRIP_RAM[$R8] = $R19
10:     $R8 = $R8++
11: **else if** $R10 == right border **then**
12:     $R21 = ($R18 + $R18)>> 1
13:     $R19 = $R19 − $R21
14:     $R21 = ($R17 + $R19 + 2) >> 2
15:     $R18 = $R18 + $R21
16:     STRIP_RAM[$R8] = $R18
17:     $R8 = $R8++
18:     STRIP_RAM[$R8] = $R19
19:     $R8 = $R8++
20: **else**
21:     $R21 = ($R18 + $R20) >> 1
22:     $R19 = $R19 − $R21
23:     $R21 = ($R17 + $R19 + 2) >> 2
24:     $R18 = $R18 + $R21
25:     $R17 = $R19
26:     STRIP_RAM[$R8] = $R18
27:     $R8 = $R8++
28:     STRIP_RAM[$R8] = $R19
29:     $R8 = $R8++
30: **end if**

---

into $R18, $R19 and $R20, where $R20 holds the pixel at location $[n+2]$. The right border check algorithm is shown in Algorithm 4.13.

Once the pixels are loaded into the DWT filter registers, the row filtering is performed as described in Algorithm 4.14. When filtering at both the left and right

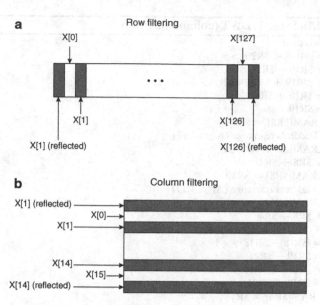

Fig. 4.14 Symmetric extension through reflection for row filtering and column filtering

---

Algorithm 4.15 Stage 1: DWT (column-filtering border check)

 1: **if** filtering at bottom border **then**
 2:     $R18 = STRIP_RAM[$R7]
 3:     $R7 = $R7 + $R4
 4:     $R19 = STRIP_RAM[$R7]
 5: **else**
 6:     $R18 = STRIP_RAM[$R7]
 7:     $R7 = $R7 + $R4
 8:     $R19 = STRIP_RAM[$R7]
 9:     $R7 = $R7 + $R4
10:     $R20 = STRIP_RAM[$R7]
11: **end if**

---

image borders, the values for the DWT filter $R17 ($[2n1]$) and $R20 ($[2n + 2]$) are not available. Therefore, to perform the row filtering as accurately as possible, a symmetrical extension is used. With the symmetrical extension, the missing pixels are reflected along the image borders leading to $R17 having the same value as $R19 for the left border case and $R20 having the value of $R18 for the right border.

Figure 4.14 shows the symmetric extension through reflection for both row- and column-filtering cases. The row-filtered coefficients are then stored according to their original pixel locations back into S3.

The column-filtering process can be carried out as soon as the row-filtering process is completed. Algorithms 4.15 and 4.16 show the border check and the column-filtering algorithms. Instead of checking for the right image border as in the row filtering, the bottom image border is checked for the column-filtering process.

Algorithm 4.16 Stage 1: DWT (column filtering)

1: **if** $R9 == 0$ **then**
2:      $R21 = (\$R18 + \$R20) >> 1$
3:      $\$R19 = \$R19 - \$R21$
4:      $\$R21 = (\$R19 + \$R19 + 2) >> 2$
5:      $\$R18 = \$R18 + \$R21$
6:      $\$R17 = \$R19$
7:      STRIP_RAM[$R8] = $R18
8:      **do** DWT address rearrange (Algo. 4.17)
9:      STRIP_RAM[$R11] = $R18
10:     $\$R8 = \$R8 + \$R4$
11:     STRIP_RAM[$R8] = $R19
12:     **do** DWT address rearrange (Algo. 4.17)
13:     STRIP_RAM[$R11] = $R19
14:     $\$R8 = \$R8 + \$R4$
15: **else if** $R9 ==$ bottom border **then**
16:     $\$R21 = (\$R18 + \$R18) >> 1$
17:     $\$R19 = \$R19 - \$R21$
18:     $\$R21 = (\$R17 + \$R19 + 2) >> 2$
19:     $\$R18 = \$R18 + \$R21$
20:     STRIP_RAM[$R8] = $R18
21:     **do** DWT address rearrange (Algo. 4.17)
22:     STRIP_RAM[$R11] = $R18
23:     $\$R8 = \$R8 + \$R4$
24:     STRIP_RAM[$R8] = $R19
25:     **do** DWT address rearrange (Algo. 4.17)
26:     STRIP_RAM[$R11] = $R19
27: **else**
28:     $\$R21 = (\$R18 + \$R20) >> 1$
29:     $\$R19 = \$R19 - \$R21$
30:     $\$R21 = (\$R17 + \$R19 + 2) >> 2$
31:     $\$R18 = \$R18 + \$R21$
32:     $\$R17 = \$R19$
33:     STRIP_RAM[$R8] = $R18
34:     **do** DWT address rearrange (Algo. 4.17)
35:     STRIP_RAM[$R11] = $R18
36:     $\$R8 = \$R8 + \$R4$
37:     STRIP_RAM[$R8] = $R19
38:     **do** DWT address rearrange (Algo. 4.17)
39:     STRIP_RAM[$R11] = $R19
40:     $\$R8 = \$R8 + \$R4$
41: **end if**

The column filtering is similar to the row-filtering process with the exception of storing the filtered DWT coefficients LL, LH, HL and HH to the STRIP_RAM using a different address generated by the new address calculation through bit rearrangement as described later in the section.

Algorithm 4.17 shows the step in which the column-filtering process uses the custom address rearranging instruction. Here, Line 2 uses the algorithm-specific instruction DWTA to obtain the new address by supplying the current address

---

Algorithm 4.17 Stage 1: DWT (DWT address rearrange)

1: $R11 = $R8 - $R12
2: $R11 = DWT_Address_Rearrange[$R11, $R6]
3: $R11 = $R11 + $R26

Algorithm 4.18 Stage 1: DWT (data and variables updates)

1: **Define:**
2: $R24 - temporary read address for LL coefficients copying
3: $R25 - temporary write address for LL coefficients copying
4: $R27 - pixel copy counter
5: **Update Variables**
6: $R3 = $R3 >> 1
7: $R4 = $R4 >> 1
8: $R5 = $R5 >> 1
9: $R5 = $R5 >> 1
10: **Check whether need to perform another round of DWT filtering**
11: $R6 = $R6 + 1
12: **if** $R6 < 5 **then**
13:     $R24 = 2048
14:     $R25 = 4096
15:     $R27 = 0
16:     **Copy LL coefficients from S2 into S3 for next level DWT**
17:     **while** $R27 < $R5 **do**
18:         $R22 = STRIP_RAM[$R24]
19:         STRIP_RAM[$R25] = $R22
20:         $R24 = $R24++
21:         $R25 = $R25++
22:         $R27 = $R27++
23:     **end while**
24:     **go to** Algo. 4.12 Line 12
25: **else**
26:     **exit** Algo. 4.12
27: **end if**

($R11) and the DWT level ($R6). If a normal MIPS computation for the new address is to be done, 14 clock cycles would be required for $l = 4$ and 65 clock cycles would be required for $l = 1$. The DWTA instruction returns the new address in one clock cycle for each filtered coefficient of DWT level $l = 1 \rightarrow 4$ and speeds up the new address computation process for Stage 1. For a l-level DWT decomposition where $l > 1$, the LL coefficients generated in the $(l - 1)$ stage are loaded from partition S2 back into partition S3. Then, a l-level decomposition (row and column filtering) is performed on these LL coefficients. The coefficient copy and parameter updates are shown in Algorithm 4.18. The DWT coefficients generated at this stage are stored back to the partition S2 using the addresses from the new address calculation.

Fig. 4.15 Arrangements of DWT coefficients in the STRIP_RAM—1D DWT. (**a**) 2-D DWT arrangement (level 3). (**b**) Desired 1-D DWT arrangement in STRIP_RAM

a

LL	LH	LH	LH	LH	LH	LH	LH
0	1	2	3	4	5	6	7
HL	HH	LH	LH	LH	LH	LH	LH
8	9	10	11	12	13	14	15
HL	HL	HH	HH	LH	LH	LH	LH
16	17	18	19	20	21	22	23
HL	HL	HH	HH	LH	LH	LH	LH
24	25	26	27	28	29	30	31
HL	HL	HL	HL	HH	HH	HH	HH
32	33	34	35	36	37	38	39
HL	HL	HL	HL	HH	HH	HH	HH
40	41	42	43	44	45	46	47
HL	HL	HL	HL	HH	HH	HH	HH
48	49	50	51	52	53	54	55
HL	HL	HL	HL	HH	HH	HH	HH
56	57	58	59	60	61	62	63

b

LL	LH	HL	HH	LH	LH	LH	LH
0	1	8	9	2	3	10	11
HL	HL	HL	HL	HH	HH	HH	HH
16	17	24	25	18	19	26	27
LH	LH	LH	LH	LH	LH	LH	LH
4	5	6	7	12	13	14	15
LH	LH	LH	LH	LH	LH	LH	LH
20	21	22	23	28	29	30	31
HL	HL	HL	HL	HL	HL	HL	HL
32	33	34	35	40	41	42	43
HL	HL	HL	HL	HL	HL	HL	HL
48	49	50	51	56	57	58	59
HH	HH	HH	HH	HH	HH	HH	HH
36	37	38	39	44	45	46	47
HH	HH	HH	HH	HH	HH	HH	HH
52	53	54	55	60	61	62	63

4.4.4.2 New Address Calculation Through Bits Rearrangement

The wavelet coefficients after Stage 1 are stored in partition S2 from addresses 2,048 to 4,095. Figure 4.15a shows an example of the wavelet coefficient arrangement in the 2D format. In the hardware implementation, the filtered coefficients are to be stored in the 1D format in the following order, LL, LH, HL and HH, as shown in Fig. 4.15b. Since the lifting-based DWT generates the coefficient outputs in mixed orders, a new address calculation is required to rearrange the output coefficients into the desired order. Figure 4.16c, d and e show the output from the lifting-based DWT for three decomposition levels and the order in which they are required to be arranged to facilitate visual processing in the next stage.

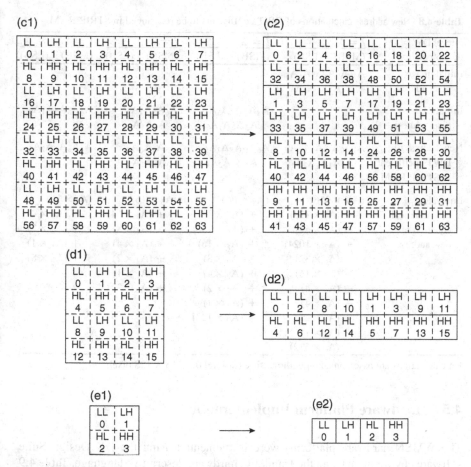

Fig. 4.16 Arrangements of DWT coefficients in the STRIP_RAM—lifting DWT. (**c1**) Lifting-based DWT filter output (level 1). (**c2**) 1-D DWT arrangement (level 1). (**d1**) Lifting-based DWT filter output (level 2). (**d2**) 1-D DWT arrangement (level 2). (**e1**) Lifting-based DWT filter output (level 3). (**e2**) 1-D DWT arrangement (level 3)

Table 4.8 shows the predefined bits arrangement used in calculating the new addresses of the DWT coefficients in the partition S2. The bit calculations are based on an input dimension of 16×128 pixels at level 0. The DWT coefficients are arranged in such a way that the processing on each sub-band can be carried out efficiently by directly loading out a continuous stream of values corresponding to each sub-band at each decomposition level. It can be seen from Table 4.8 that the new address calculation equations provide the new addresses through the rearrangement of the original address bits. This will only require simple rewiring and does not increase the hardware complexity significantly.

Table 4.8 New address calculations of DWT coefficients to be rearranged in STRIP_RAM

l-level DWT decomposition	$l=1$ (MSB) (LSB)	$l=2$ (MSB) (LSB)	$l=3$ (MSB) (LSB)	$l=4$ (MSB) (LSB)
Initial image pixel address	$A_{10}A_9A_8A_7A_6A_5$ $A_4A_3A_2A_1A_0$	–	–	–
Initial LL pixel address	–	$A_8A_7A_6A_5A_4$ $A_3A_2A_1A_0$	$A_6A_5A_4A_3$ $A_2A_1A_0$	$A_4A_3A_2A_1A_0$
New address of DWT coefficient in STRIP_RAM	$A_7A_0A_{10}A_9A_8A_6$ $A_5A_4A_3A_2A_1$	$A_6A_0A_8A_7A_5$ $A_4A_3A_2A_1$	$A_5A_0A_6A_4$ $A_3A_2A_1$	$A_4A_0A_3A_2A_1$
Equivalent mathematical equation of new address	$(A_{10} \times 256)$ $+ (A_9 \times 128)$ $+ (A_8 \times 64)$ $+ (A_7 \times 1024)$ $+ (A_6 \times 32)$ $+ (A_5 \times 16)$ $+ (A_4 \times 8)$ $+ (A_3 \times 4)$ $+ (A_2 \times 2)$ $+ (A_1 \times 1)$ $+ (A_0 \times 512)$	$(A_8 \times 64)$ $+ (A_7 \times 32)$ $+ (A_6 \times 256)$ $+ (A_5 \times 16)$ $+ (A_4 \times 8)$ $+ (A_3 \times 4)$ $+ (A_2 \times 2)$ $+ (A_1 \times 1)$ $+ (A_0 \times 128)$	$(A_6 \times 16)$ $+ (A_5 \times 64)$ $+ (A_4 \times 8)$ $+ (A_3 \times 4)$ $+ (A_2 \times 2)$ $+ (A_1 \times 1)$ $+ (A_0 \times 32)$	$(A_4 \times 16)$ $+ (A_3 \times 4)$ $+ (A_2 \times 2)$ $+ (A_1 \times 1)$ $+ (A_0 \times 8)$

Bit calculations are based on an input dimension (at level 0) of 16×128 pixels

4.5　Hardware Platform Implementations

The WMSN hardware platforms were implemented using the DK Design Suite software environment using the Handel-C hardware descriptive language. Table 4.9 shows a summary of the platform implementation for the low-cost RC10 implementation. The implementation includes the soft-core processor, global clock and controls, memory blocks and the I/O interfaces between the FPGA, the main board, the camera port and the RS-232 port. The FPGA power summary was generated using the Xilinx XPower Estimator 11.1 [142] software for Spartan-3 FPGAs and gave an estimated output power of 224 mW.

4.6　Chapter Summary

This chapter has presented the WMSN hardware platforms and its implementation on the Celoxica FPGA development boards. The platform contains the soft-core processor for visual processing (strip-based MIPS). The strip-based MIPS processor was customised with pipelining, forwarding and hazard detection hardware and an

Table 4.9 Hardware platform summary for Celoxica RC10 WMSN hardware platform

Imager	
Camera	Omnivision OV9650
	1.3 megapixel colour CMOS camera
Image capture	QVGA (320 × 240), RGB
Capture modes	Normal or low light
Capture output	128 × 128, greyscale approximation
	(RGB green component)
Power	50 mW
Main board, soft-core processors and memory	
Board	Celoxica RC10 development board
FPGA device	Xilinx Spartan-3 XC3S1500L
I/O ports	Camera connector, RS-232 port
Processor	Strip-based MIPS, URISC AES
Global clock	37.5 MHz
Instruction memory	$2,048 \times 16$ bits
Data memory	$64 \times 128 \times 16$ bits (STRIP_RAM)
Frame memory	$128 \times 128 \times 16$ bits (IMAGE_RAM)
Power	224 mW
Wireless Module	
End device	Digi XBee Pro Series 2
Operating frequency band	ISM 2.4 GHz
Baud rate	9,600 symbols/s
Power	2 mW

extended instruction set to increase its throughput. The following chapters will make use of the WMSN hardware platform for various visual processing applications focusing on single-view and multiple-view information reduction techniques.

Chapter 5
Single-View Information Reduction Techniques for WMSN Using Event Detection

Abstract This chapter discusses single-view information reduction techniques using event detection for the WMSN hardware platform. In this information reduction method, the captured image is partitioned into a number of patches. Each patch is examined to determine if it contains useful events in the scene. Patches which do not contain useful events are not transmitted. The event detection is performed using a visual attention (VA) model. The chapter begins by giving some background on event detection approaches and then moves on to describe VA and its use in vision processing applications. Next, a review of VA algorithms and architectures is given, and a specific VA model based on medium spatial frequencies (MSF) is selected for implementation. The MSF model is modified to reduce computational and memory requirements for the hardware implementation. The implementation of the MSF VA model on the strip-based MIPS processor is then discussed and concludes with the implementation of the VA information reduction technique on the WMSN hardware platform.

5.1 Introduction

Event detection approaches reduce scene data by only transmitting image frames when significant events are detected for the application. For example, a surveillance application could use a face detector to decide which image frames to send to the base station. However, the face detector would need to have low computational complexity to meet the energy requirements in the WMSN. There is a trade-off between the energy required for processing and the energy required for transmission. On the one hand, using an event detector in the WMSN requires more computational power. On the other hand, this could result in a saving of transmission power when frames are discarded. The other advantage of an event detector is that it could also serve as an early stage for visual pre-processing. To perform the facial recognition process, the central computer would need to perform at least two stages.

The first stage is to locate the face location in the image, and the second stage would then perform the recognition task by comparing the facial features with a stored database. To reduce the large amount of image data for processing by the central computer, the event detector performs the face detection task and the location of the face is then communicated to the central computer to perform the facial recognition task. An example of a WMSN employing an event detector can be found in the paper by [37]. The authors propose an event detector using simple image processing at the camera nodes based on difference frames and the chi-squared detector. Their algorithm performs well on indoor surveillance sequences and some outdoor sequences. Event detection is a mature field in computer vision. A comprehensive review of event detection can be found in the book by [55]. The research challenge is to find suitable detectors which are reliable and can be efficiently implemented within the hardware constraints of the WMSN.

Computational visual attention (VA) can be defined as algorithmic models that provide a means of selecting specific parts of the visual input information for further higher vision processing or investigation based on principles of human selective attention [44]. Research on the underlying mechanism of VA began in the 1890s where psychological studies on the nature of the attentive process were carried out [69]. For a visual scene seen by an observer, attention will cause the allocation of processing resources to salient areas rather than the entire scene as a strategy to work around the limited processing capacity of the human brain. As a result, parts of the visual scene can be broken down into a series of localised and computationally less demanding information for rapid processing. The purpose of attention is to shorten the reaction time for the human visual system (HVS) to process a visual scene by first attending to salient areas which would bring about a visual awareness and understanding of the entire scene [68, 108]. Two advantages of VA applied to vision systems are described in [44]. First, the handling of complex visual data in detection- and recognition-related tasks can be simplified. Conventional detection and recognition approaches require the use of multiple sliding window sizes and trained classifiers. With VA processing, priority image regions can be identified and used for detection and recognition tasks without any prior training. The second advantage for the use of VA in vision processing is the support for action decisions. Vision systems often have to operate in complex and unknown environments. It is necessary for the system to locate and interpret relevant parts of the visual scene to decide on the actions to be taken. VA allows such relevant parts to be easily detected based on the modelled mechanism of attention found in the HVS.

The following list provides some examples of vision processing and applications in which VA has been used:

- *Artificial vision in robotics.*
  An artificial vision allowing feature detection, tracking and active camera control to be performed in a mobile robot was developed using a visual simultaneous localisation and mapping (SLAM) system based on attentional landmarks [45]. The visual SLAM system had to detect and track landmarks that are conspicuous and stable over several frames as well as map-visited locations through the

redetection of those landmarks. A VA model was utilised for the detection of conspicuous landmarks which allowed the visual SLAM to focus on sparse landmark representations.

- *Image compression.*
  Image compression aided with VA processing allows conspicuous regions to be compressed with a lossless compression algorithm, while the background and non-conspicuous regions are compressed with a lossy compression algorithm. This form of adaptive compression retains the quality of the important information and achieves a high compression ratio. Examples can be found in the works of [74] and [99].

- *Object recognition [39].*
  A Bayesian network is incorporated into a VA model for object recognition. The Bayesian network for selected objects from the input image is trained with the maximum feature location from each feature map, and the selected objects are represented with a probability distribution of the most salient location found in the feature maps. When a location in an input image is specified, the trained model will list the probabilities showing the likelihood of known objects found in the specified location.

- *Object segmentation and extraction.*
  The input image undergoes a VA processing and an image segmentation to obtain a saliency map and segmented region map, respectively. Objects present in the image can then be extracted by matching the salient locations indicated by the saliency map to the segmented regions. The VA processing allows quick identification of conspicuous objects for extraction without the need for an object database. Examples can be found in the works of [89] and [52].

- *Video compression.*
  A region-of-interest (ROI) coding for video compression based on the multi-resolution property of the phase spectrum of quaternion Fourier transform (PQFT) has been proposed by [51]. The PQFT was used to compute spatiotemporal saliency maps for ROI selection in video sequences. Using the PQFT VA model, ROIs consisting of conspicuous objects related to the features of colour, luminance and motion were detected over multiple wavelet decomposition scales. The detected ROIs served as masks which were applied to each decomposition level. Video compression was performed by retaining all the coefficients in the ROI masks, while the coefficients below the bit plane threshold (BPT) of the non-masked regions were omitted.

While VAs offer an artificial visual attention mechanism that can be applied towards a wide range of vision processing algorithms, it has a disadvantage in terms of computational complexity. The complex algorithm structures of VA models have to be taken into consideration when implementing in an environment with a limited computational capability. The computational time of the VA processing has to be evaluated to meet the requirements for real-time applications. Furthermore, blob-like detected conspicuous regions seen in early VA models are being out-phased by higher resolution outputs in newer models resulting in an increase of computational cost and memory requirements. This makes the implementation of VA models in resource constrained environments such as the WMSN particularly challenging.

5.2 Computational Visual Attention and Models

This section will briefly review some concepts in computational visual attention (VA) such as high- and low-level features, bottom-up and top-down approaches, centre-surround process and saliency map. The section will also briefly review VA models and algorithms which have been proposed in the literature.

5.2.1 *High-Level and Low-Level Features*

Features present in a visual stimuli can be divided into two categories: high-level features and low-level features. High-level features allow for a semantic bridge between the extracted visual information and the interpretation by the user resulting in the understanding of the entire visual scene. These features exist in various subjective forms and are often defined by data or image representations. While high-level features provide a more accurate and semantic understanding of the visual scene, the extraction of these features is time consuming and involves large training databases [75]. In contrast to high-level features, the distinctive attributes for low-level features can be easily extracted from objects or regions in a visual scene. However, low-level features contribute less to the understanding of the visual scene. Although it is unable to provide a full semantic understanding of the entire visual scene, low-level features play an important role in providing the quality of being unique for the information, locations and objects present in the visual stimuli. Examples of low-level features which a human can detect easily are colour, edge, luminance, orientation, scale, shape and texture.

5.2.2 *Bottom-Up and Top-Down Approaches*

Visual attention can be computed using two different approaches: bottom-up approach and top-down approach. The top-down approach is task driven and is related to recognition processing. It is more complex to model since it integrates high-level processes such as task, cognitive state and visual context. For the bottom-up approach, attention is driven by the low-level features present in the visual stimuli. The approach can be said to be task independent since the observer would have no prior knowledge of the visual stimuli or what actions should be performed. Figure 5.1 shows examples of visual salience caused by low-level

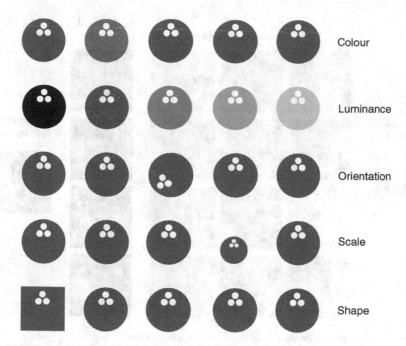

Fig. 5.1 Examples of visual salience caused by low-level features

features of colour, luminance, orientation, scale, shape and texture. An observer viewing the five objects for each row would experience a redirection of attention towards the visually salient object. Then the observer would grasp an understanding of the object's nature and the property that makes the object salient. This is caused by the bottom-up computation of salience by the HVS to first redirect attention to the visually salient object in order for further analysis to obtain an understanding of the visual stimuli. The bottom-up VA approach is closely related to early vision where visual features are extracted and processed in the retina layer and some parts of the cortical layer V1 [61]. The top-down VA approach can be associated with the higher vision where task specific operations are performed in the cortical layer V2, V3 and V4.

5.2.3 Centre-Surround Process

The centre-surround (CS) process is an operation that resembles the preference of stimuli cells found in the lateral geniculate nucleus (LGN) of the visual pathway [124]. In VA, the CS process generally functions to provide an enhancement of fine feature details found at higher resolutions and coarse feature details at lower resolutions prior to the combination of multiple features that leads to a global

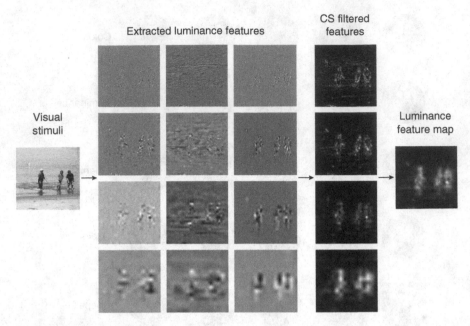

Fig. 5.2 Centre-surround process using the sliding window method

contribution of salience. The CS process can be computed in various ways with the same goal of enhancing locations consisting of both fine and coarse details from their surroundings indicated by the CS differences or also known as the contrast differences. Among the methods used for computing the CS differences are the point-to-point subtraction of locations between fine and coarse scales [67], the use of a sliding window [1] and the use of morphological operators [125]. Figure 5.2 illustrates the CS process using the sliding window method.

5.2.4 Saliency Map

The saliency map is the end result of the VA processing performed in a bottom-up approach. The concept of a saliency map was introduced to provide a measure of visual salience from selected contributing features present in visual stimuli. The saliency map can be seen as a topographical map containing information from the available features in the visual stimuli with the aim of providing a global measure of salience. The saliency map can be used as a control mechanism for selective attention [71], for vision tasks such as the prediction of eye movements [43] and for computer vision applications involving content-based information retrieval and object segmentations [81]. Figure 5.3 shows an input image for VA processing and its saliency map where the global salience is computed from the colour,

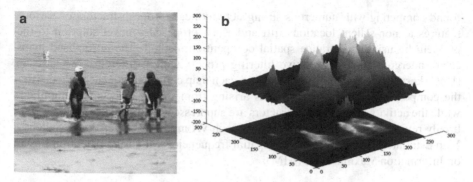

Fig. 5.3 Input image for VA processing and its saliency map

luminance and orientation features. The peaks in the surface plot of the saliency map indicate locations of high salience giving a representation of conspicuous locations in the visual stimuli.

5.2.5 Visual Attention Models

Many VA models use the bottom-up approach and are based on the feature integration theory (FIT) of Treisman and Gelade [124]. The FIT describes the pre-attentive stage of the visual system where separable features present in the visual stimuli can be detected and identified in parallel. The authors also described attention with a spotlight metaphor whereby the attention of the observer moves around the field of vision like a spotlight beam. Based on the FIT, Koch and Ullman developed a computational and neurally plausible architecture for controlling visual attention [71]. In this model, the low-level features extracted and computed in parallel are stored as a set of cortical topological maps called feature maps. The feature maps are then combined into a single saliency map that encodes feature conspicuity at each location in the visual stimuli. A winner-takes-all (WTA) strategy is utilised to allow competition within the computed maps to localise the most active location in the saliency map.

Using the framework of the Koch and Ullman's VA model, Itti et al. introduced a newer implementation of a VA model for rapid scene analysis [67]. This model uses linear filtering and centre-surround structures in the feature computations and allows low-level features consisting of colour, luminance and orientation to be computed quickly in parallel across 42 spatial maps. Later on, Itti and Koch improved on the Itti et al. VA model by further investigating the feature combination structure across several modalities [66]. From observations, they discovered that the attention system suffers from a severe signal-to-noise ratio problem when a large number of maps are used in the combination process. The locations of salient objects consisting of strong activity peaks arising from a few distinctive features of the objects are

·found competing with numerous strong activity peaks from distractors of various features at non-salient locations. Itti and Koch proposed a novel solution to the problem by using an iterative spatial competition method. The normalised maps each undergo a 10-stage iterative filtering process with a difference-of-Gaussian (DoG) kernel where activity peaks within each map compete with each other. After the competition process, strong peaks arising from salient objects are amplified, while the activity peaks from distractors are suppressed. In the next section, we will briefly review four VA models: Walther–Koch VA model, Wavelet-Based Saliency Map Estimator (WBSME), medium spatial frequencies (MSF) and Attention Based on Information Maximisation (AIM).

5.2.5.1 Walther–Koch Visual Attention Model

The Walther–Koch VA model [133] is an extension of the Itti et al. [67] and the Itti–Koch [66] VA models which are commonly used for visual attention comparisons. The input image is first repeatedly subsampled by means of convolution and decimation by a factor of two into a nine-scale dyadic Gaussian pyramid. A linearly separable Gaussian kernel of $[1\,5\,10\,10\,5\,1]/32$ is used in creating the image pyramid. In the Gaussian pyramid hierarchy, the image at level zero corresponds to the input image, and the image at the peak of the pyramid corresponds to the subsampled image at level eight. The intensity map is computed as

$$M_I = \frac{r + g + b}{3} \tag{5.1}$$

where the channels of the RGB image are r (red), g (green) and b (blue). This operation is repeated for all levels in the image pyramid forming an intensity pyramid $M_I(\sigma)$. For the colour features, two opponent colour pyramids for red–green (RG) and blue–yellow (BY) are created using Eqs. (5.2) and (5.3):

$$M_{RG} = \frac{r - g}{\max(r, g, b)} \tag{5.2}$$

$$M_{BY} = \frac{b - \min(r, g)}{\max(r, g, b)} \tag{5.3}$$

To avoid fluctuations of colour opponency at low luminance values, M_{RG} and M_{BY} are set to zero at pixel locations with $\max(r, g, b) < 1/10$ for a dynamic range of $[0, 1]$. The orientation maps are computed by convolving the levels of the intensity pyramid with Gabor filters as shown in Eqs. (5.4) and (5.5). The results of the convolution operations are four maps of orientations $0°$, $45°$, $90°$ and $135°$;

$$O(\sigma) = \|I(\sigma) * G_0(\theta)\| + \|I(\sigma) * G_{n/2}(\theta)\| \tag{5.4}$$

where

$$G_\psi(x, y, \theta) = \exp\left(-\frac{x'^2 + \gamma^2 y'^2}{2\delta^2}\right) \cos\left(2\pi\frac{x'}{\lambda} + \psi\right) \tag{5.5}$$

is a Gabor filter with an aspect ratio γ, wavelength λ, standard deviation δ and phase ψ. The transformed coordinates (x', y') with respect to orientation θ are given by Eqs. (5.6) and (5.7) as

$$x' = x\cos(\theta) + y\sin(\theta) \tag{5.6}$$

$$y' = -x\sin(\theta) + y\cos(\theta) \tag{5.7}$$

The algorithm uses the values of $\gamma = 1$, $\lambda = 7$ pixels, $\delta = 7/3$ pixels and $\psi \in \{0, \pi/2\}$. The filters are truncated to 19×19 pixels. After the visual features have been extracted and grouped into their feature pyramids, a CS structure is applied to the pyramids to compute a set of feature maps. The CS operations are implemented using across level subtraction ($\ominus$) between two maps at the centre (c) and the surround (s) scales of each feature pyramid. The centre corresponds to a pixel at level $c \in 2, 3, 4$ of the pyramid, and the surround corresponds to a pixel at level $s = c + \alpha$ where $\alpha \in 3, 4$. The CS operation for the intensity feature is given as

$$I(c, s) = |I(c) \ominus I(s)| \tag{5.8}$$

The colour feature maps are computed in the same manner as for the intensity feature maps in Eq. (5.8). For the orientation feature maps, the CS process is computed for all four orientations as

$$O(c, s, \theta) = |O(c, \theta) \ominus O(s, \theta)| \tag{5.9}$$

where $\theta \in \{0°, 45°, 90°, 135°\}$. In total, there will be 42 feature maps computed: six for intensity, 12 for colour and 24 for orientation. All the feature maps are normalised using the normalisation operator $N(.)$ to globally promote maps that have strong peaks of activities and to suppress maps with numerous comparable peak responses. The normalisation operator is given as

$$N(.) = \frac{I - I_{min}}{I_{max} - I_{min}} \times 255 \tag{5.10}$$

where is the lowest value pixel in the image and is the highest value pixel in the image. Next, the feature maps are summed according to individual feature using the across-scale addition ($\oplus$), and the sums are normalised once more to form the conspicuity maps:

$$\bar{I} = \bigoplus_{c=2}^{4} \bigoplus_{s=c+3}^{c+4} N(I(c,s)) \tag{5.11}$$

$$\bar{C} = \bigoplus_{c=2}^{4} \bigoplus_{s=c+3}^{c+4} [N(RG(c,s)) + N(BY(c,s))] \tag{5.12}$$

$$\bar{O} = \sum_{\theta=\{0°,45°,90°,135°\}} N\left(\bigoplus_{c=2}^{4} \bigoplus_{s=c+3}^{c+4} N(O(c,s,\theta))\right) \tag{5.13}$$

Finally, the three conspicuity maps are normalised and summed. The sum of these maps is averaged, giving a single saliency map:

$$S = \frac{1}{3}\left(N(\bar{I}) + N(\bar{C}) + N(\bar{O})\right) \tag{5.14}$$

5.2.5.2   Wavelet-Based Saliency Map Estimator (WBSME) Model

While the Itti–Koch model [66] performs well for detecting locations of conspicuous objects in a given image, its overall process is computationally lengthy due to the need to extract and group visual features into their respective pyramids before feature maps can be computed. Noticing the inefficiency of computing feature maps in the Itti–Koch model, Tsapatsoulis et al. proposed an alternative model for computing visual salience while maintaining the biological framework of the model termed the Wavelet-Based Saliency Map Estimator (WBSME) model [126]. In this model, the Gaussian pyramid is replaced with a multi-resolution pyramid created through a series of wavelet filtering and subsampling operations. The discrete wavelet transform (DWT) is a viable alternative to the Gaussian pyramid as the HVS appears to process visual input in a multi-resolution manner similar to the wavelet transform [20].

The WBSME takes in a YC_bC_r image as input. To perform the DWT at the start, a pair of low-pass and high-pass filters is applied to the input image channels Y, C_b and C_r in both the horizontal and vertical directions. The filter outputs are then subsampled by a factor of two. For one level of wavelet decomposition, there will be four sub-bands: LL, HL, LH and HH. The LL sub-band contains the approximate coefficients and is a coarser representation of the input image. The LL sub-band of the Y channel is used to compute the intensity feature whereas the LL sub-bands of the C_b and C_r channels are used to compute the colour features. The other three sub-bands containing the HL (vertical detail coefficients, LH (horizontal detail coefficients) and HH (diagonal detail coefficients) will be used for the computation of the orientation features. The next step of the algorithm consists of the CS operation where the CS differences are computed at level j using a point-to-point subtraction between the approximation at the next coarser level $(j + 1)$ with the approximation at the current j-th level.

The feature maps at various levels j are then combined to compute the conspicuity maps for the intensity, colour and orientation features. The conspicuity maps are computed by interpolating the feature maps to a finer level, summed using a point-to-point addition and then passed through a saturation function. The saturation function is a more accurate way of combining features from different modalities compared to the normalisation, summation and averaging method used in the Itti and Koch model which creates inaccurate results as it weakens the importance of salient peaks from individual modalities. The end result is an intensity conspicuity map C_I. The same steps are also applied to the colour and orientation features to obtain the colour conspicuity map, C_C, and the orientation conspicuity map, C_O. The final saliency map, S, is obtained as

$$S = \frac{2}{1 + e^{-(C_I + C_C + C_O)}} - 1 \qquad (5.15)$$

5.2.5.3 Medium Spatial Frequencies (MSF) Salience Model

Urban et al. proposed an attention model termed the medium spatial frequencies (MSF) model based on the framework of Koch and Ullman that operates on medium- to high-frequency features to predict the best fixation locations in free-viewing tasks [129]. Similar to the WBSME VA model, the MSF model utilises wavelet decomposition for early feature extraction. A window CS filtering method is used in place of the point-to-point subtractions used in the fine and coarse scale approach. To validate the biological plausibility of their proposed model, the authors conducted an eye fixation experiment to allow a comparison of fixation points to the saliency maps generated. From the eye fixation evaluations, the MSF model was able to provide high predictability of the fixated eye locations while outperforming several other VA models. The fixation locations are best predicted with medium to high frequencies for man-made scenes and low to medium frequencies for natural scenes.

The MSF model extracts early visual features by means of the wavelet transform using the YC_bC_r colour space. Each individual Y, C_b and C_r channel is decomposed using the 9/7 Cohen–Daubechies–Feauveau (CDF) discrete wavelet transform (DWT). For each channel, a five-level decomposition is performed. As medium spatial frequencies are of interest, only the orientation sub-bands HL, LH and HH are retained for processing. The LL sub-band is discarded. After the five-level feature sub-bands for each channel have been computed, a CS process is applied to each orientation sub-band for the three channels. This gives the orientation maps. For each location, the CS is computed as the difference between the current pixel coefficient value and the mean coefficient value of the neighbouring pixels as shown in Eq. (5.16):

$$CS(x) = \left| |I(x)| - \frac{1}{s} \sum_{k \in S} |I(k)| \right|$$ (5.16)

where $I(x)$ is the pixel coefficient value at location x, S is the centre-surround support and s is the surround area. The surround area for this model is fixed to the size of 5×5 pixels for each pyramid level. The next step of the algorithm consists of summing up the orientation maps according to pyramid level and channel to form level feature maps. For each level, a filtering process is applied as shown in Eq. (5.17):

$$L(x) = \frac{1}{d} \sum_{k \in D(x)} \left(\sum_{o \in (1,2,3)} CS_o(k) \right)$$ (5.17)

where $CS_o(k)$ is the centre-surround response at location k for the orientation sub-bands, $D(x)$ is a disc of radius $1°$ centred on k and d is the surface area of the disc. The final step of the algorithm is the fusion of the level maps. The fusion step consists of two stages: level fusion and channel fusion. The level fusion step is applied first by averaging the sum of level maps performed through successive bilinear upsampling and point-to-point addition to give channel maps as shown in Eq. (5.18):

$$C(x) = \frac{1}{Nb\_L} \oplus L_l(x)$$ (5.18)

where Nb_L is the number of decomposition level and $L_l(x)$ is the map at level l. The saliency map is then computed by summing all three channel maps and averaging them (channel fusion). Finally, the saliency map is normalised to the range of $[0, 255]$. Equation (5.19) shows the computation of the saliency map $S(x)$:

$$S(x) = N \left(\frac{1}{3} \left(C_Y(x) + C_{C_b}(x) + C_{C_r}(x) \right) \right)$$ (5.19)

where $C_Y(x)$, $C_{C_b}(x)$ and $C_{C_r}(x)$ are the channel maps for channels Y, C_b and C_r, respectively.

5.2.5.4   Attention Based on Information Maximisation (AIM) Model

Bruce and Tsotsos proposed a different computational VA model termed the Attention Based on Information Maximisation (AIM) model [16]. This model is based on self-information where the model predicts salience based on spatial locations that carry the most information. The AIM model consists of three main stages: independent feature extraction, density estimation and joint likelihood and self-information. In the independent feature extraction stage, each pixel location $I(m, n)$ of the input image is computed for the response to a set of various basis functions B_k. To compute the basis functions B_k, 31×31 and 21×21 RGB patches are sampled from a set of 3,600 natural scene images picked randomly from the Corel Stock Photo Database. A joint approximate diagonalization of eigenmatrices (JADE)-based independent component analysis (ICA) [19] is then applied to these patches, retaining a variance of 95 % with principal component analysis (PCA) as a pre-processing step. The ICA results in a set of basis components where mutual dependence of features is minimised. The retention of 95 % variance yields 54 basis filters for patches of 31×31 and 25 basis filters for patches of 21×21. The basis function which is also called the mixing matrix is then pseudoinversed. Pseudoinversing the mixing matrix provides an unmixing matrix which can be used to separate the information within any local region of the input into independent components. The matrix product between the local neighbourhood of the pixel and the unmixing matrix gives the independent feature of an image pixel as a set of basis coefficients. This output of the independent feature extraction stage is shown in Eq. (5.20):

$$a_{m,n,k} = P * C_k \qquad (5.20)$$

where $a_{m,n,k}$ is the set of basis coefficients for the pixel at location $I(m, n)$, P is the unmixing matrix computed by the pseudoinverse of the basis functions $B_{m,n,k}$ for location (m, n), $*$ is the linear product between two matrices and C_k is a 21×21 local neighbourhood centred at (m, n). The next stage involves determining the amount of information carried by a pixel. The information may correspond to features such as edges and contrast and is measured in terms of the coefficient densities of the pixel. For a local neighbourhood $C_{m,n,k}$ at pixel location (m, n), the content of this neighbourhood corresponding to the image I is characterised by a number of basis coefficients a_k resulting from the matrix multiplication with the unmixing matrix. Each basis coefficient a_k can be related to a basis filter.

Surrounding this local neighbourhood, another larger neighbourhood $S_{m,n,k}$ exists with $C_{m,n,k}$ at its centre. The size of S_k is determined to span over the entire image I. The surrounding neighbourhood $S_{m,n,k}$ also possesses a number of basis coefficients a_k that are also possessed by $C_{m,n,k}$. In order to estimate the coefficient density or likelihood of a pixel at (m, n), basis coefficients corresponding to the same filter type in spatial locations of both $C_{m,n,k}$ and $S_{m,n,k}$ are observed, and distributions of these coefficients are computed by means of histogram density estimates. A joint likelihood of all the coefficients within a neighbourhood C_k that corresponds to a pixel location is found from the product of all individual likelihoods. The computation of the joint likelihood for a particular pixel is assumed to have non-dependency between individual likelihoods due to the sparse representation of the ICA. Hence, each coefficient density contributes independently to the joint likelihood of a location situated in the image I. Equation (5.21) shows the computation of joint likelihood for a particular pixel $I(m, n)$:

$$P\left(I(m, n)\right) = \prod_{k=1}^{N_B} P\left(C_{m,n,k}\right) \qquad (5.21)$$

where $P\left(I(m, n)\right)$ is the joint likelihood of pixel $I(m, n)$, $P\left(C_{m,n,k}\right)$ is the coefficient density corresponding to the basis function k and N_B is the total number of basis function. The joint likelihood is then translated as shown in Eq. (5.22) to give the self-information map:

$$-\log\left(P\left(I(m, n)\right)\right) = \sum_{k=1}^{N_B} -\log\left(P\left(C_{m,n,k}\right)\right) \qquad (5.22)$$

The self-information map is then convolved with a Gaussian envelope that corresponds to an observed drop-off in visual acuity. This encodes the amount of information gained for a target location. The self-information map represents the salience of spatial locations in the input image. Locations that have higher self-information are regarded to contain more information and are considered more salient.

5.2.5.5   Comparison of VA Models

This section will give a comparison of the Walther–Koch, MSF and AIM VA models for two types of scenes: simple and complex. The simple scene images were taken from the Berkeley Segmentation Dataset [87], and the complex scene images were taken from the MIT Indoor Scene Recognition Database [105]. Five images for the simple and complex scenes were picked from the datasets. Prior to the VA processing, all test images were resized to the dimensions of 256×256. As the saliency maps generated by the Walther–Koch model were smaller in dimension compared to their inputs, the maps of this model were upsampled by means of bilinear interpolation for the visual evaluations.

Simple Scenes

Figure 5.4 shows the saliency maps for the three VA models using simple scene images. For the image "Airplane", all three VA models failed to detect the aircraft as a whole. However, the locations that an observer would be likely to fixate upon were predicted correctly to a certain extent. The Walther–Koch model managed to predict the locations correctly with the exception of the right wing. The WBSME and MSF saliency maps highlighted the edges where the boundary between the

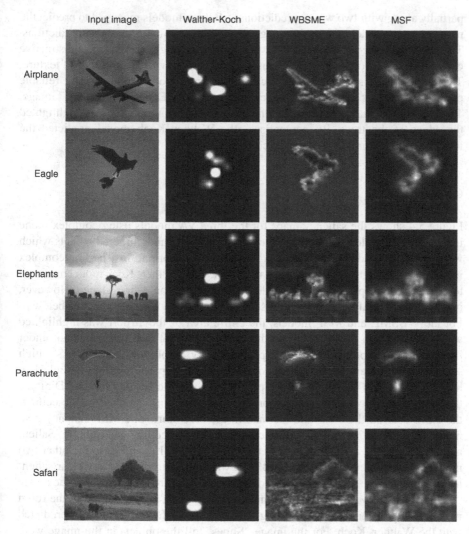

Fig. 5.4 Visual comparison of saliency maps for VA models using simple scene images

aircraft and the sky occurred, covering the tail, wings and the nose but failed to predict the fuselage. In both cases, more salient points were found concentrated on the tail and the right wing. For the second image "Eagle", the location which a viewer would fixate upon besides the eagle itself would be the tail which is different in colour compared to its body. In this case, all three models predicted the location correctly. For the image "Elephants", the WBSME and the MSF models managed to predict all the locations consisting of the elephants and the tree. However, noises arising from texture of the background were detected and amplified in the saliency map of WBSME. The Walther–Koch model only managed to detect the elephants partially along with two wrong predictions. All three models managed to predict the parachute and parachutist in the image "Parachute" without any false predictions. This shows that the three VA models work well with images consisting of distinctive objects and background. However, noises are amplified with the presence of texture. The amplified noises can render the saliency map from providing any reliable predictions. This is seen in the saliency map of WBSME using the "Safari" image. The MSF model managed to predict the location of the animal but only highlighted the edges of the trees. By comparison, the Walther–Koch model predicted the locations of the animal and the trees correctly.

Complex Scenes

Figure 5.4 shows the saliency maps for the three VA models using complex scene images. The complex scene images consist of multiple and cluttered objects which were harder to predict without the aid of top-down information. The first complex image consists of groups of people in an airport. The Walther–Koch model managed to predict several people which would draw attention on a first glance. However, it also made several false predictions such as the balcony area and the area below it. For the WBSME and MSF models, the entire area of the airport was highlighted with the exception of the floors. In the saliency map of WBSME, the spot under illumination was pointed to by the peak saliency amplitude. For the MSF, high saliency amplitudes indicated locations where the apparel worn by people had high contrast with the floor. In the prediction outcome for the image "Dining", the Walther–Koch model correctly predicted the lighting attached to the ceiling, the black coat on the chair and the man in the blue shirt. The MSF saliency map highlighted the entire visual scene except for the floor, ceiling and lighting. Salient amplitudes peaked at the locations of the three women. In contrast to the other two models, the saliency map for the WBSME was unable to provide any clear salient locations due to the map being affected by noise arising from textures. For the image "Playroom", the WBSME and MSF managed to predict all the objects in the room including the lightings on the ceiling whereas only a few objects were predicted with the Walther–Koch. For the image "Shoes", all the objects in the image were predicted by the WBSME and MSF models with the same salient peaks (pointing to the reflection in the bottom right mirror). The Walther–Koch only detected the shoes that were of high contrasts (red, blue, yellow shades) and the reflection in the left mirror. For this particular image the WBSME predictions were the most accurate compared to the other two models with the locations corresponding to the objects having clear boundaries. In the final image "Subway", all three models managed to predict the locations of the people sitting on the bench. The region covering the people and the bench can be made out from the saliency maps for WBSME and MSF. The WBSME again displayed a noisy saliency map due to the textured background.

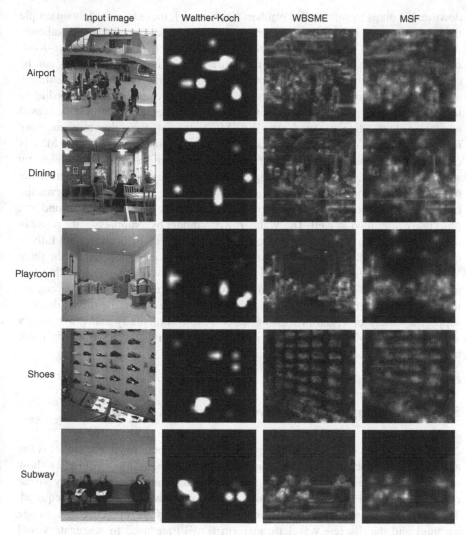

Fig. 5.5 Visual comparison of saliency maps for VA models using complex scene images

Evaluation of Saliency Maps

Figures 5.4 and 5.5 show that all three VA models are able to predict the conspicuous locations containing important or interesting objects accurately to a certain extent. The Walther–Koch VA model encodes conspicuous locations in a blob-like manner where locations are highlighted based on importance instead of the objects themselves. The blob size also determines the coverage area of the high amplitude salient pixels. The saliency maps for the WBSME and MSF VA models are very much similar. Instead of encoding conspicuous locations in downscaled maps as seen in the Walther–Koch model, these two models upsample the maps operated upon at each combination stage. Because of this, the saliency maps are higher in resolution compared to the saliency map of the Walther–Koch model. One advantage of this is that the boundary of the predicted objects can be easily established making the conspicuous locations useful for object segmentation applications [2]. Furthermore, the conspicuous locations can be ranked according to individual pixel amplitude instead of region-based values. However, the increased resolution of the saliency maps for the WBSME and the MSF requires more memory resources than the Walther–Koch model. In terms of noise suppression, the MSF is capable of suppressing noises arising from illumination and texture considerably well. The WBSME has low noise suppression due to the method of computing the orientation sub-bands for the orientation features. While the method preserves the edges enclosing the objects in the input image, textures from the background and objects are amplified as well. The Walther–Koch model oversuppresses the noise in the input, leading to high region compactness at conspicuous locations, but fails to maintain the boundaries of the objects. From the evaluations conducted for the three VA models, it is evident that the MSF has the overall advantage over the other two models. While the Walther–Koch and the WBSME each has their own strengths, they are more suited for specific applications rather than a versatile model that can support a variety of visual processing tasks. A low-complexity and low-memory adaptation of the MSF VA model for implementation on the WMSN platform will be discussed next.

5.3 Low-Complexity and Low-Memory VA Model

An important consideration when implementing a computational VA model is the amount of embedded memory which is available on chip. To model closely to how neurons analyse visual stimuli over multiple spatial scales in the primary visual areas of the human eye, a multi-resolution image processing technique is required. Typically multi-resolution processing techniques such as the Gaussian image pyramid and the discrete wavelet transform (DWT) are used to segregate visual features spread over a range of spatial scales into multiple image maps. Before these image maps can be used for feature extraction, they have to be stored into memory. Due to the need to retain the image and feature maps at every multi-resolution level, memory requirements would be high. A single high-resolution image could easily occupy several kilobytes of RAM. The high-memory occupancy of computational VA models makes them very challenging for implementation in embedded hardware. Another issue regarding the implementation of VA models is that many of them require floating-point arithmetic. This makes them difficult to implement and also to achieve real-time processing speed in embedded systems.

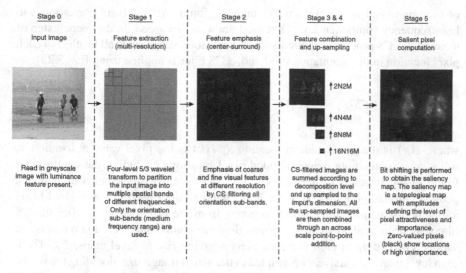

Fig. 5.6 An overview of the integer-based MSF

5.3.1 Integer-Based MSF Model for Hardware Implementation

An integer-based model for the MSF is proposed in this section replacing its floating-point computations to reduce the computational complexity for implementation in hardware. The proposed model retains the major stages of the original MSF model. The modifications of the algorithm are performed in such a way that all required multiplications and divisions operations are of the order of 2^i where $i = 0, 1, 2, \ldots$ so that these operations can be implemented as bit shift instructions. Furthermore, a strip-based processing approach is applied in the modified model to reduce the memory requirements. The modified VA model will only operate based on a single luminance feature to reduce the number of VA processing by 67 %. The modified low-complexity and low-memory VA model is then evaluated in terms of prediction performance and memory requirements in comparison to its original algorithm. Figure 5.6 shows the overview of the proposed integer-based VA model.

5.3.1.1 Integer-Based Computations

The initial step of the model is to extract visual features from the image. The desired visual features of interest can be found in different orientation ranges contained in medium- to high-frequency ranges. In order to decompose the image to several sets of orientations relating to a certain frequency range, the 5/3 DWT is applied to the image using a four-level decomposition. The final decomposition would result in four levels with each level consisting of three orientation sub-bands: horizontal, vertical and diagonal. The final level will have a sub-band which is an approximation of the image and four times smaller than the input. As this band corresponds to low-frequency components, it is not used and it is discarded. In the second step of the model, a CS process is applied to each orientation sub-band at all levels. For each pixel location in the orientation sub-band, a CS filter is applied using Eq. (5.23):

$$CS(x) = \left| |I(x)| - \frac{1}{s} \left(4 \cdot |I(x)| + \sum_{k \in NE} |I(k)| \right) \right| \qquad (5.23)$$

where $I(x)$ is the pixel value at location x, $I(k)$ is the pixel value at location k, NE is a set containing neighbourhood pixel locations excluding the centre pixel found in the structured filter and s is the total number of elements used in the pixel neighbourhood. Instead of the 5×5 window filter used in the original MSF, a diamond-structured filter with two pixels from the origin is used. After the CS filter has been applied to all the orientation sub-bands, the filtered sub-bands are summed according to their respective levels giving rise to level maps, L_p. Then, each level map is recursively upsampled to the actual image size, doubling the image dimension for each recursion as described in Eq. (5.24):

$$L = (L_p) \uparrow_l^{2N2M} \qquad (5.24)$$

where L is the upsampled level map; $\uparrow_l^{2N2M}$ denotes up-sampling by a factor of two along the rows and columns, respectively; and l is the number of recursion relating to the number of decomposition level of the operated L_p. The final step consists in fusing the upsampled level maps. The fusion operation is performed by means of point-to-point addition, $\oplus$, of pixels across all level maps. The channel map, C, is then obtained from Eq. (5.25):

$$C(x) = \frac{1}{ND} \bigoplus_l L_l(x) \qquad (5.25)$$

where ND is the total number of wavelet decomposition used and L_l is the up-sampled level map at level l. When multiple features are used, the saliency map, S, is the average of the sum of channel maps from each feature. For simplicity in referring to the MSF variants, the integer-based MSF and the original MSF will be termed as MSF 5/3 and MSF 9/7, respectively.

5.3.1.2 Diamond-Structured Centre-Surround Filter

Figure 5.7 illustrates the diamond-structured CS filter used in the feature emphasis stage of the MSF 5/3 VA model. The CS filtering is conducted in a snake-like horizontal manner. The CS filter is proposed to overcome the need for floating-point computations used in the original 5×5 window CS filter found in the MSF 9/7 VA algorithm. The centre location of the proposed filter is given a weighting of

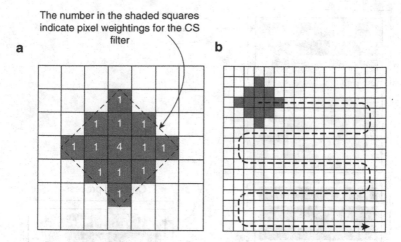

Fig. 5.7 Diamond-structured filter used in the CS process of the MSF 5/3 VA model

four, while its neighbouring locations are given a weighting of one. The structure of the proposed CS filter provides two advantages. The first advantage is to allow the bit shift operation to be used as a substitute to the floating-point division. Instead of having to divide by 25, the neighbourhood averaging now requires a division by 16 which can be performed by four right bit shifts. This leads to simplification of the hardware complexity. The second advantage is that the proposed CS structure allows for a certain degree of noise suppression. Locations that have sharp amplitude peaks in the form of isolated pixels are often seen originating from the noises present in the visual input. By giving the centre location of the CS filter a higher weighting, the CS operation would give a suppressed CS value for the centre location which is desirable for a high precision saliency output. In the case where sharp contrasts occur at locations that contain important information, the CS differences of the neighbouring location will compensate for the suppressed value of the centre location.

5.3.2 Lowering Memory Requirements Through Strip-Based Processing

The strip-based processing is an approach which allows high-resolution images as well as imaging methods that require high-memory requirements to be performed in an environment with limited memory resources. The main idea of the approach is to partition the input image into multiple even-sized strips and having each strip processed individually through an iterative loop. The even-sized strips are referred to as image strips, and each strip is treated as a valid input for processing. With the approach, the amount of memory required for processing can be lowered

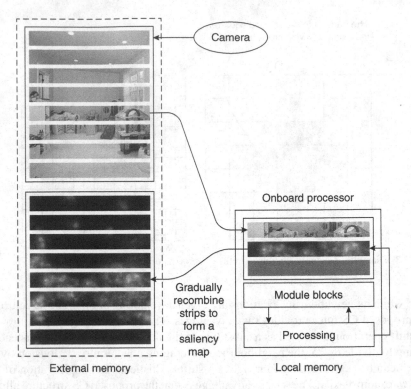

Fig. 5.8 Strip-based processing approach applied to VA model

according to the number of image strips partitioned from the input image. Although a minor increase in instruction overhead is expected due to the iteration, the advantage of the strip-based processing is attractive for implementing VA in a resource constrained environment. Figure 5.8 shows the strip-based processing approach applied to the MSF 5/3 VA model. The image captured by the camera is first stored into the external memory. After the image capture has been obtained, a segment of the full image in strip form is loaded from the external memory into the smaller local memory for VA processing. When the VA processing has been completed, the processed strip is released and written back to the external memory for recombination. At the end of the iterative strip-based approach, a fully recombined saliency map will be in the external memory. In practical cases, the saliency strip is kept in the local memory at each iteration to be used with the task-dependent processing of the system. Once the task-dependent processing has been completed, the final resultant strip will be either written back to the external memory for recombination or directly transmitted to the base station before the next iteration begins.

Table 5.1 Memory requirements against the number of strips for the MSF 5/3 implementation

Number of strips (n)	Number of locations	Memory required (kB)	Memory reduction (%)
1	65,536	131,072	0
2	32,768	65,536	50
4	16,384	32,768	75
8	8,192	16,384	87.5

5.3.2.1 Memory Reduction for the MSF 5/3

For the MSF 5/3 VA algorithm, a four-level wavelet decomposition is used in the feature extraction stage. This in turn causes the remaining stages to work with maps of four levels. With a carefully designed architecture, the MSF 5/3 VA can be implemented using only two memory arrays which will be referred to as IMAGE_RAM and STRIP_RAM. The IMAGE_RAM is the external memory of the system and is used to store the image captured by the camera. The STRIP_RAM is the smaller local memory. The STRIP_RAM is used to hold the data involved in the VA processing as well as other data from tasks that are performed by the on-board processor. For a single feature used (the RGB green component in this case), the IMAGE_RAM holds a greyscale image. The size of the IMAGE_RAM required is then equal to the dimension of the greyscale image capture as shown in Eq. (5.26):

$$IMAGE\_RAM\_size = N \times M \tag{5.26}$$

where N and M are the number of rows and columns in the captured greyscale image, respectively. Considering the scenario where the computed saliency strip is used for additional task-dependent processing and the image strip has to be retained in the STRIP_RAM for further use, the size of the STRIP_RAM has to be four times larger than the IMAGE_RAM. While the IMAGE_RAM may be used to fill in the inadequate amount of local memory, situations may arise where the IMAGE_RAM itself is barely sufficient to sustain an effective processing in the system. However, with the strip-based processing approach, the size of the STRIP_RAM can be varied according to the number of strips n assigned in the hardware architecture as described by Eq. (5.27):

$$STRIP\_RAM\_size = 4 \times (IMAGE\_RAM\_size/n) \tag{5.27}$$

Table 5.1 shows the required memory for the MSF 5/3 for $n = 1, 2, 4, 8$. The calculations are performed using an input image of dimension 128×128 with 16 bits (2 bytes) allocated to each location in the STRIP_RAM. It can be observed that the memory requirement for the integer-based MSF implementation in hardware decreases significantly with the increase in the number of strips used.

Fig. 5.9 Visual comparison of saliency maps for various strip sizes

5.3.2.2   Comparison of Saliency Maps for Various Strip Sizes

Figure 5.9 shows a visual comparison of saliency maps with the proposed MSF 5/3 for various strip sizes with $n = 1, 2, 4, 8$. For the saliency maps obtained through the strip-based processing, most of the predicted salient points are similar to the ones in the saliency maps without the strip-based processing applied. However, it can be seen that there are horizontal line artefacts present in the computed saliency maps. The problem occurs due to the loss of information at the image strip borders during the feature extraction stage. With the use of the strip-based processing, clippings at the image strip borders are observed. By doubling the image strip row dimension $(2n)$ and using only the centre portion of the image strip corresponding to the number of rows after the wavelet decomposition in the feature extraction stage, the problem of the line artefacts can be eliminated. However, this method comes with a trade-off in terms of memory requirement and requires twice as much memory. The number of falsely predicted salient points can be reduced by comparing the local maximum and minimum of the image strip which is under VA processing to the local maximum and minimum of the prior-processed image strip. The largest maximum and the smallest minimum from the comparison is then used to normalise the image strip under processing. While the normalisation accuracy increases progressively leading to the reduction in false positives, additional hardware resources are required for the comparison. While the saliency maps computed using the strip-based processing may seem distorted when observed visually, the distortions do not hinder the salience predictions from providing reliable predictions although a drop in the overall performance score and the relevance to human eye fixation is expected as a trade-off to the high reduction in memory.

5.4 Hardware Implementation on the WMSN Platform

This section describes the implementation of the MSF 5/3 VA model on the WMSN hardware platform. The implementation consists of six stages: data copy, discrete wavelet transform (DWT), feature emphasis, feature combination, upsampling and across level combination and salient pixel computation. In the implementation, these algorithms are translated and implemented as a set of MIPS binary instructions.

5.4.1 Stages 0 and 1: Data Copy and DWT

The hardware implementations for Stages 0 and 1 (Data Copy and DWT) have been described in Sects. 4.4.3 and 4.4.4, respectively.

5.4.2 Stage 2: Feature Emphasis

Algorithm 5.1 shows the structure of the feature emphasis stage. The DWT coefficients from Stage 1 have been arranged in a one-dimensional (1D) stream in the following order: LL4, LH4, HL4, HH4, LH3, HL3, ...LH1, HL1 and HH1.

Algorithm 5.1 Stage 2: Feature emphasis

1: **Define:**
2: $R3 - sub-band LH, HL, HH row size
3: $R4 - sub-band LH, HL, HH column size
4: $R5 - number of pixels in current sub-band
5: $R6 - start DWT level l, Stage 2 counts down from $l = 4$ to $l = 1$
6: $R7 - read address pointer (reads from S2), addresses 2048 - 2055 is LL sub-band
7: $R8 - write address pointer (writes to S2)
8: $R9 - row counter
9: $R10 - column counter
10: $R11 - write address pointer (writes to S3)
11: $R16 - sub-band counter
12: **Initialisations:** $R3 = 1; $R4 = 8; $R5 = 8; $R6 = 4; $R7 = 2056; $R8 = 2056; $R16=3
13: **Initialisations for padding:** $R0 = 0; $R10 = 0; $R11 = 4096
14: **while** $R6 != 0 **do**
15:    **while** $R16 != 0 **do**
16:       **do** padding through pixel replication (Algorithm 5.2)
17:       **do** center-surround operation (Algorithm 5.4)
18:       **Reset parameters for next sub-band operation**
19:       $R9 = 0
20:       $R10 = 0
21:       $R11 = 4096
22:       $R16 = $R16−−
23:    **end while**
24:    **Update parameters for next dwt level operations**
25:    $R3 = $R3 << 1
26:    $R4 = $R4 << 1
27:    $R6 = $R6−−
28:    $R11 = 4096
29:    $R16 = 3
30: **end while**

To read these coefficients and write back the CS output to the respective locations in the same order, two conditional while loops are used. The outer loop (Line 14) cycles through the DWT levels, starting from $l = 4$ down to $l = 1$. The inner loop (Line 15) cycles through the sub-bands of DWT level l defined by the outer loop in the order of LH, HL and HH. Within the inner loop, padding of the sub-band image is performed and followed by the CS operation. The 1D stream of the DWT coefficients loaded from the STRIP_RAM has to be padded prior to the CS filtering operation. For the padding process, the LH, HL and HH DWT coefficients from partition S2 are loaded and padded before being written to S3. During this process, if a coefficient is found to lie at the image borders, two pixels are replicated and added to the 1D data write stream. If a line of coefficients corresponding to an entire row is found lying at the top or bottom of the 2D image layout, then two similar valued lines are replicated for each of the conditions, respectively. The padded coefficients are stored in the 1D arrangement in S3 before they are CS filtered.

Algorithm 5.2 Stage 2: Feature emphasis (padding through pixel replication)

1: $R13 = $R4 − 1
2: **while** $R9 < $R3 **do**
3:    **while** $R10 < $R4 **do**
4:        $R12 = STRIP_RAM[$R7]
5:        **Left and right borders padding**
6:        **if** ($R10 == 0 || $R10 == $R13) **then**
7:            STRIP_RAM[$R11] = $R12
8:            $R11 = $R11++
9:            STRIP_RAM[$R11] = $R12
10:           $R11 = $R11++
11:           STRIP_RAM[$R11] = $R12
12:           $R11 = $R11++
13:       **else**
14:           STRIP_RAM[$R11] = $R12
15:           $R11 = $R11++
16:       **end if**
17:       $R7 = $R7++
18:       $R10 = $R10++
19:    **end while**
20:    **Top border padding**
21:    **if** $R9 == 0 **then**
22:        **do** double row line padding (Algorithm 5.3)
23:    **end if**
24:    $R9 = $R9++
25:    $R10 = 0
26: **end while**
27: **Bottom border padding**
28: **do** double row line padding (Algorithm 5.3)

Algorithm 5.2 describes the steps of the padding through pixel replication. Line 1 first assigns a right border compare value to $R13. This allows the algorithm to check for a right-hand border pixel by comparing the current column counter $R10 with the value stored in $R13. Lines 2 and 3 each defines the row and column loops to cycle through the sub-band for the padding operation. The DWT LH coefficient at $l = 4$ pointed to by the S2 read address $R7 is loaded to a temporary register $R12. Then $R10 is checked in Line 6 whether the current column lies on the left or the right border. If it lies on the left or the right border, the value in $R12 is written to S3 with the address defined by $R11, and $R11 is incremented by one. This write and increment operation is repeated two more times. If the condition indicates the current column does not lie on the left or right borders, the write and increment operation is only performed once (Lines 14–15). The column loop iterates with the S2 read address incremented by one at the end of every loop until the last column is reached.

Once the column loop exits, the current row is checked for the top row condition. If the current row is a top row, then the top border padding is performed using Algorithm 5.3, else it loops back to the start of the row loop. Algorithm 5.3 shows the steps for a double row line padding. This pads two rows identical to the row that

---

Algorithm 5.3 Stage 2: Feature emphasis (double row line padding)

```
 1: $R25 = 0
 2: $R23 = $R4 + 4
 3: $R24 = $R11 − $R23
 4: $R23 = $R23 << 2
 5: while $R25 < $R23 do
 6:     $R12 = STRIP_RAM[$R24]
 7:     $R24 = $R24++
 8:     STRIP_RAM[$R11] = $R12
 9:     $R11 = $R11++
10:     $R25 = $R25++
11: end while
```
---

have just been left or right padded and written to S3. Line 1 of this algorithm first defines a padding loop counter $R25 initialised to zero. $R23 in Line 2 is given a padded column value (12 in the case of $l = 4$). A padding start read address $R24 is generated by subtracting the current S3 write address $R11 with the value in $R23. This brings the padding read address back to the location of the first pixel in the previous row that has just been processed. Lines 5–11 then duplicate the processed row twice into S3. When the row loop exits, Algorithm 5.3 is applied again to pad the bottom border.

Once the padding through pixel replication has been performed, Algorithm 5.4 is then used to CS filter the padded data. The first step of the CS algorithm is to compute the start address for the CS filter whereby its origin is located at coordinate $(2, 2)$ if the padded coefficients are viewed as a single 2D image. To compute the filter start address, the equation given in Line 9 is used. The column size is first added by four to give the padded column size (additional 2 pixels on each image border after padding). Then, the sum is left bit shifted once to shift the row downwards by one and added by two, giving the location after the padded left border values. Finally, the CS filter correct start address is determined by adding an offset of 4,096 (start address of STRIP_RAM S3) to the result. These steps are shown in Lines 11–14. A description of the filter start address computation for padded coefficients from sub-bands of level $l = 1$ is found in Fig. 5.10.

When the start address for the filter has been determined, the CS operation is carried out, cycling through the number of coefficients contained in the sub-band that is operated upon. From this point, the coefficients are loosely called as pixels since the sub-bands are treated as images. Within the loops of Lines 16 to 17, the centre pixel, O, bounded by the filter is first loaded into register $R12. The loaded centre pixel value is then absolute valued and stored into the weighted centre value register $R17. Subsequently, the absolute value is left bit shifted twice to give a weighting of four, and this weighted value is stored to the neighbourhood pixel sum register $R18. Then, the neighbourhood pixels A to L as shown in Algorithm 5.5 are visited, processed and added to the neighbourhood sum in $R18. Once all the pixels

---

Algorithm 5.4 Stage 2: Feature emphasis (centre-surround operation)

1: **Define:**
2:   $R11 - start address of STRIP_RAM partition S3
3:   $R17 - weighted center value
4:   $R18 - neighbourhood pixel sum
5:   $R26 - CS STRIP_RAM read pointer
6:   $R27 - CS row counter
7:   $R28 - CS column counter
8: **Initialisations:** $R11 = 4096; $R17 = 0; $R18 = 0; $R27 = 0; $R28 = 0
9: **Require:** Start read address = 2 × (column size + 4) + 2 + offset
10: **Compute CS start read address**
11: $R14 = $R4 + 4
12: $R23 = $R14 << 1
13: $R26 = $R26 + 2
14: $R26 = $R26 + $R11
15: **Center-surround operation**
16: **while** $R27 < $R3 **do**
17:    **while** $R28 < $R4 **do**
18:       **Center pixel, O**
19:       $R12 = STRIP_RAM[$R26]
20:       $R17 = | $R12 |
21:       $R18 = $R17 << 1
22:       $R18 = $R18 << 1
23:       **Load, process and sum neighbourhood pixels**
24:       **do** process neighbour pixel contained in filter (Algorithm 5.5)
25:       **Compute difference between center pixel and neighbourhood**
26:       $R18 = $R18 >> 1
27:       $R18 = $R18 >> 1
28:       $R18 = $R18 >> 1
29:       $R18 = $R18 >> 1
30:       $R18 = | $R17 - $R18 |
31:       STRIP_RAM[$R8] = $R18
32:       $R8 = $R8++
33:       $R26 = $R26++
34:       $R28 = $R28++
35:    **end while**
36:    $R26 = $R26 + 4
37:    $R27 = $R27++
38:    $R28 = 0
39: **end while**

---

in the filter neighbourhood are visited, the CS difference is computed by absoluting the difference between the value of $R17 and the average of the $R18 value. The CS difference is then written back to STRIP_RAM S2, to the same location where the pixel O was initially loaded out for padding. The CS process is repeated until the LH, HL and HH sub-bands of $l = 4$ down to $l = 1$ have been processed. At the end of Stage 2, the CS filter pixels have replaced the coefficients of Stage 1 in S2.

Algorithm 5.5 Stage 2: Feature emphasis (process neighbour pixels)

1: **Neighbour pixel, B**
2: $R19 = R26 - R14$
3: **do** process neighbour pixel (Algorithm 5.6)
4: **Neighbour pixel, A**
5: $R19 = R19 - R14$
6: **do** process neighbour pixel (Algorithm 5.6)
7: **Neighbour pixel, E**
8: $R19 = R19 + R14$
9: $R19 = R19 - 1$
10: **do** process neighbour pixel (Algorithm 5.6)
11: **Neighbour pixel, F**
12: $R19 = R19 + 2$
13: **do** process neighbour pixel (Algorithm 5.6)
14: **Neighbour pixel, H**
15: $R19 = R26$
16: $R19 = R19 - 1$
17: **do** process neighbour pixel (Algorithm 5.6)
18: **Neighbour pixel, G**
19: $R19 = R19 - 1$
20: **do** process neighbour pixel (Algorithm 5.6)
21: **Neighbour pixel, I**
22: $R19 = R19 + 3$
23: **do** process neighbour pixel (Algorithm 5.6)
24: **Neighbour pixel, J**
25: $R19 = R19 + 1$
26: **do** process neighbour pixel (Algorithm 5.6)
27: **Neighbour pixel, C**
28: $R19 = R26 + R14$
29: **do** process neighbour pixel (Algorithm 5.6)
30: **Neighbour pixel, K**
31: $R19 = R19 - 1$
32: **do** process neighbour pixel (Algorithm 5.6)
33: **Neighbour pixel, L**
34: $R19 = R19 + 2$
35: **do** process neighbour pixel (Algorithm 5.6)
36: **Neighbour pixel, D**
37: $R19 = R19 - 1$
38: $R19 = R26 + R14$
39: **do** process neighbour pixel (Algorithm 5.6)

Algorithm 5.6 Stage 2: Feature emphasis (process individual neighbour pixel)

1: $R12 = STRIP\_RAM[R19]$
2: $R17 = |\ R12\ |$
3: $R18 = R18 + R12$

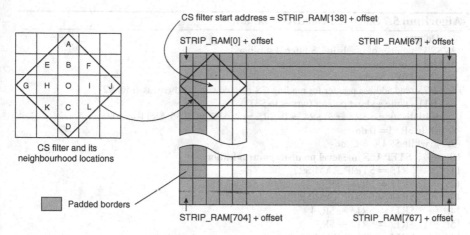

Fig. 5.10 Computing the CS filter start address

5.4.3   Stage 3: Feature Combination

The feature combination of Stage 3 involves the point-to-point addition of the same level CS-filtered orientation sub-band maps. From the CS operation in the previous stage, the data in S2 consists of the following order: LL4, LH4, HL4, HH4, LH3, HL3, HH3, LH2, HL2, HH2, LH1, HL1 and HH1. All the orientation sub-bands have been CS filtered except for the approximate sub-band LL4 which would be discarded. The feature combination algorithm begins by processing the CS-filtered coefficients of the level $l = 4$ orientation sub-bands and then progresses in a decrementing order down to $l = 1$. The structure of Stage 3 is shown in Algorithm 5.7. From Lines 11 to 21, the same level point-to-point addition is shown. The CS-filtered coefficient of LH4 at location 2,056 is first loaded out to $R13. Then the S2 read address is incremented by the total number of pixels in the $l = 4$ sub-band to the first CS-filtered coefficient of HL4 at location 2,064 where this value is added to $R13. The read address is incremented and the value at 2,072 is added to $R13. Now $R13 holds the combined feature value of the first location of the same level map at level 4. This value is written to S3 at location 4,096. The process of adding the CS-filtered coefficients of the three sub-bands LH, HL and HH is repeated until level maps of level 4 to 1 are obtained.

When the point-to-point addition has completed, S3 holds the same level maps in the order of level 4 down to level 1 at addresses 4,096–4,775. As the next stage (Stage 4) requires S3 for temporary storage and computations as well have its results stored in S2, the same level map pixel values of Stage 3 are copied to S4 occupying the locations 6,144–6,824. In this case, the same level maps are stored with the level 1 map first and the level 4 map last. The algorithm of the level map copy is shown in Algorithm 5.8. Finally, S2 is cleared of its data by setting the locations 2,048–4,095 to zero as shown in Algorithm 5.9. The overall process of Stage 3 is shown in Fig. 5.11.

Algorithm 5.7 Stage 3: Feature combination

1: **Define:**
2: $R3 - number of pixels in CS filtered sub-band
3: $R4 - pixel counter
4: $R5 - DWT level l
5: $R7 - read address pointer for reading CS filtered sub-bands(reads from S2)
6: $R11 - write address pointer (writes to S3)
7: **Initialisations:** $R3 = 8; $R4 = 0; $R5 = 4; $R7 = 2056; $R11 = 4096
8: **while** $R5 != 0 **do**
9:      **while** $R4 < $R3 **do**
10:         **STEP 1: Same level point-to-point addition**
11:         $R13 = STRIP_RAM[$R7]
12:         $R12 = $R7 + $R3
13:         $R14 = STRIP_RAM[$R12]
14:         $R13 = $R13 + $R14
15:         $R12 = $R12 + $R3
16:         $R14 = STRIP_RAM[$R12]
17:         $R13 = $R13 + $R14
18:         STRIP_RAM[$R11] = $R13
19:         $R7 = $R7++
20:         $R11 = $R11++
21:         $R4 = $R4++
22:     **end while**
23:     $R7 = $R7 + $R3
24:     $R7 = $R7 + $R3
25:     $R3 = $R3 << 1
26:     $R3 = $R3 << 1
27:     $R4 = 0
28:     $R5 = $R5--
29: **end while**
30: **STEP 2: Point-to-point addition result copy to S4**
31: **Initialisation:** $R11 = 6144
32: **Level 1 Copy initialisations:** $R3 = 512; $R4 = 0; $R7 = 4264
33: **do** point-to-point addition result copy (Algorithm 5.8)
34: **Level 2 Copy initialisations:** $R3 = 128; $R4 = 0; $R7 = 4136
35: **do** point-to-point addition result copy (Algorithm 5.8)
36: **Level 3 Copy initialisations:** $R3 = 32; $R4 = 0; $R7 = 4104
37: **do** point-to-point addition result copy (Algorithm 5.8)
38: **Level 4 Copy initialisations:** $R3 = 8; $R4 = 0; $R7 = 4096
39: **do** point-to-point addition result copy (Algorithm 5.8)
40: **STEP 3: Clear partition S2**
41: **do** clear partition S2 values to zero (Algorithm 5.9)

5.4.4   Stage 4: Upsampling and Across Level Combination

The same level maps from level $l = 1$ to $l = 4$ of Stage 3 are upsampled to the dimension of the image strip (16×128) in the algorithm of Stage 4. In most VA models, the bilinear interpolation is used to upsample the respective level maps. However, floating-point computations are not supported in the hardware implementation leading to the difficulty of implementing the bilinear interpolation accurately. The next approach is to use a four neighbourhood averaging (FNA)

Algorithm 5.8 Stage 3: Feature combination (data copy)

```
1: while $R4 << $R3 do
2:     $R13 = STRIP_RAM[$R7]
3:     $R7 = $R7++
4:     STRIP_RAM[$R11] = $R13
5:     $R11 = $R11++
6:     $R4 = $R4++
7: end while
```

Algorithm 5.9 Stage 3: Feature combination (data clear)

```
1:  Define:
2:  $R3 - write address pointer (writes to S2)
3:  $R5 - number of pixels in partition S2
4:  $R6 - pixel counter
5:  Initialisation: $3 = 2048; $R5 = 2048; $R6 = 0
6:  while $R6 << $R5 do
7:      STRIP_RAM[$R3] = 0
8:      $R3 = $R3++
9:      $R6 = $R6++
10: end while
```

upsampling whereby the missing pixels are found by averaging their surrounding four neighbour pixels. Figure 5.12 shows the upsampling method that uses the FNA. Consider an example input of dimension 2×2 seen in Fig. 5.11a. The input is upsampled using the FNA to give the output shown in Fig. 5.12b. The alphabetical order in Fig. 5.12a,b represents the order in which the pixel values are read and written, respectively. The line address read and write pointers indicate the start locations of separate addressing used in reading from the STRIP_RAM S4 and writing to S3. The columns of the output are ordered in the order of even–odd where the first column, $col = 0$, is an even index. Figure 5.12c shows the seven steps of the FNA upsampling to arrive at the output of (b). The first step involves filling in the left border. Then, the second step fills the output matrix from left to right in a checker board fashion. Pixel locations A to D are first filled, while E and F are found through the four neighbourhood averaging. When four neighbour pixels are not available, the missing neighbours are reflected along the computed locations with each reflected neighbour indicated by a circle. Step 3 continues to fill the bottom row with alternating locations. Note that Step 1 and Step 2 do not fill the bottom row. The output matrix after Step 3 can be seen as a full checker board pattern. Steps 4–7 fill the remaining missing pixels in terms of even and odd columns. Steps 4 and 5 fill the odd columns starting from the top of the matrix until the second last row. Similarly, Steps 6 and 7 fill the even columns starting from the second row until the final row. At the end of Step 7, a fully upsampled image is obtained.

Although the FNA upsampling method is simple and straightforward, the upsampled pixel values are rather accurate with the exception of the values at the borders where reflected pixels are required. For the dimension of the image strip, any inaccuracy in the upsampled borders is not noticeable as they are only a pixel wide.

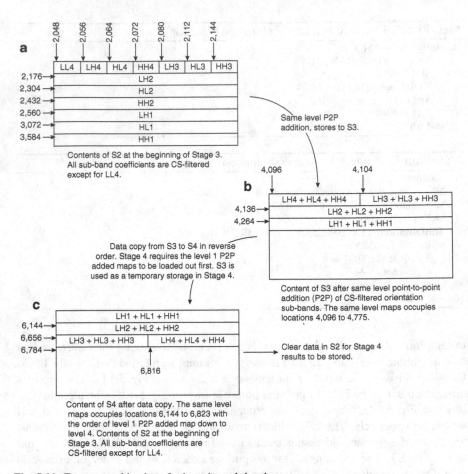

Fig. 5.11 Feature combination of orientation sub-band maps

Algorithm 5.10 shows the main structure of the Stage 4 upsampling and across level combination algorithm. In contrast to the decomposition level looping in Stages 2 and 3, Stage 4 starts off with level 1 and progresses up to level 4. The decomposition level looping is performed by the most outer loop of the Stage 4 algorithm, and it bounds the steps within Lines 11–37 in Algorithm 5.10. With the FNA upsampling method, doubling in both row and column dimensions can only be done once at a time due to its simple computational structure. For example, a $l = 4$ level map has to be upsampled iteratively four times in order for it to reach the dimension of 16×128.

Line 13 first clears the number of iterations for the upsampling required to achieve the dimension of 16×128 with the row and column counters reset in Lines 14 and 15. Then, the number of iterations required for a particular level l is computed through repeated division of the row size of the image strip with the level map row size. Once the number of upsampling iteration has been found,

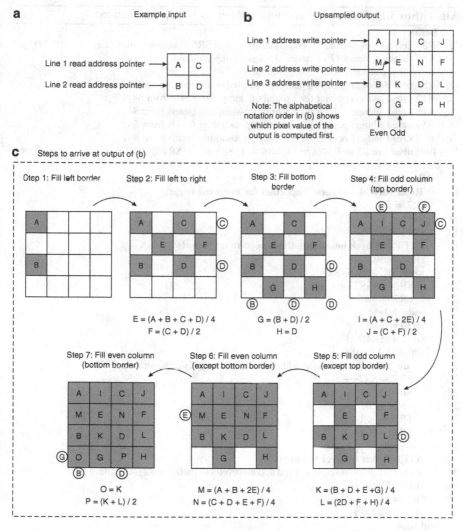

Fig. 5.12 Four neighbourhood averaging in the upsampling method

Algorithm 5.11 is used to determine the starting line read and write addresses for the upsampling process as shown in Algorithm 5.12. Then, the upsampling process is performed with either one of the two upsampling algorithms: nearest neighbour interpolation (Algorithm 5.13) or FNA (Algorithm 5.14). When the level map is found to be of $l = 4$, the nearest neighbour interpolation is applied due to the map's dimension of 1×8, else the FNA upsampling method is applied.

For the nearest neighbour interpolation, the 1×8 level map to be upsampled is interpolated to 2×16 by replicating each pixel by one to the right and by one to the bottom. For the FNA upsampling, the output matrix located in S3 is filled up in a checker board fashion. Algorithms 5.15, 5.16 and 5.17 show the detailed steps of

---

Algorithm 5.10 Stage 4: Upsampling and across level combination

1: **Define:**
2: $R3 - current row size; $R4 - current column size; $R5 - current column size
3: $R6 - level counter; $R7 - number of up-sampling required
4: $R11 - row counter; $R12 - column counter
5: $R8 - row size of up-sampled data (up-sampled once from the current)
6: $R17 - read address pointer for 1 round up-sampling (reads from S4)
7: $R18 - read address pointer for 2 rounds up-sampling (reads from S4)
8: $R19 - read address pointer for 3 rounds up-sampling (reads from S4)
9: $R20 - read address pointer for 4 rounds up-sampling (reads from S4)
10: **Initialisation:** $3 = 8; $R4 = 64; $R6 = 1; $R8 = 16; $R16; $R17 = 6144; $R18 = 6656;
     $R19 = 6774; $R20 = 6816
11: **while** $R6 < $R5 **do**
12:      **Reset variables to zero (important for repeated loops)**
13:      $R7 = 0
14:      $R11 = 0
15:      $R12 = 0
16:      **STEP 1: Check how many times need to up-sample**
17:      $R23 = $R3
18:      **while** $R23 != $R8 **do**
19:          $R23 = $R23 << 1
20:          $R7 = $R7++
21:      **end while**
22:      $R30 = $R3
23:      $R31 = $R4
24:      **while** $R7 != 0 **do**
25:          **do** determine read write addresses and up-sampling (Algorithm 5.11)
26:          $R7 = $R7--
27:          $R11 = 0
28:          $R12 = 0
29:          $R30 << 1
30:          $R31 << 1
31:      **end while**
32:      **STEP 4: Across level combination**
33:      **do** across level combination and update results to partition S2 (Algorithm 5.23)
34:      $R6 = $R6++
35:      $R3 = $R3 >> 1
36:      $R4 = $R4 >> 1
37: **end while**

Step 1 (fill left border), Step 2 (fill left to right) and Step 3 (fill bottom row) seen in Fig. 5.11c.

After the checker board fashion output matrix is obtained, the remaining missing pixels can be computed. These remaining missing pixels are found using the algorithm structure described in Algorithm 5.18. Filling up the remaining pixels of the checker board pattern consists of two main steps: filling up the odd columns and filling up the even columns. The odd columns are chosen to be filled first since they can be filled from the top row and also provide more neighbouring points to be used later on in the even column filling process. Algorithm 5.18 begins the pixel filling by looping through the rows and detecting odd columns. If the current row

---

Algorithm 5.11 Stage 4: Upsampling and across level combination (intermediate steps)

1: **Compute new row and column size of 1 round up-sampling**
2: $R9 = $R30 << 1$
3: $R10 = $R31 << 1$
4: **STEP 2: Determine read and write addresses according to up-sampling rounds**
5: **do** get read and write addresses for **STEP 3** (Algorithm 5.12)
6: **STEP 3: Begin up-sampling**
7: **if** $R7 == 4$ **then**
8:    **do** nearest neighbour interpolation (Algorithm 5.13)
9: **else**
10:    **do** four neighbours averaging (Algorithm 5.14)
11: **end if**

---

counter $R11$ is found to have a value of zero, Algorithm 5.19 is used to fill up the first row of the output matrix, else Algorithm 5.20 is used to fill up the missing pixels of the odd columns within the third row to the second last row of the output matrix.

After all the odd columns are filled, the rows of the output matrix are looped once again, scanning for even columns to be filled. Algorithm 5.22 will begin computing the missing pixels starting from the second row. When the last row is reached, Algorithm 5.21 will fill up the missing pixels for that row and complete an iteration of the upsampling process. In Algorithm 5.19, the top border of the output matrix is filled starting from the second column to the last column in column increment of twos. Other than the last column, the columns of the top borders are filled by averaging four neighbours where the missing fourth neighbour (north) is a reflection of the neighbour in the second row (south). For the last column, the missing pixel is computed by averaging the sum of the available west and south neighbours.

For the odd column filling described by Algorithm 5.20, the missing pixels are found by averaging four available neighbours except for the last column (right image border). In this case, only three neighbours are available with the east neighbour pixel missing. Therefore, the west neighbour is reflected to the east giving the missing neighbour the same pixel value as the west neighbour. Algorithm 5.22 describes the filling in of missing pixels in even columns with the exception of the bottom border. Here, the missing pixels are found by averaging the four available neighbours. The left border is an exception due to a missing left neighbour. The missing neighbour is obtained by reflecting the east neighbour with the location of the missing pixel to be computed at the centre.

The final step of the upsampling involves filling up the bottom border as described in Algorithm 5.21. The missing pixel of the first column is found by averaging the sum of neighbours at the north and at the east. The rest of the even columns of the bottom border are computed similar to the top border. The FNA upsampling method can be observed to have a diagonal neighbour averaging pattern (NW, NE, SW, SE) for Steps 1–3 and an upright cross averaging pattern (N, S, E, W) for Steps 4–7. The simplified equations shown in Fig. 5.12c are

---

Algorithm 5.12 Stage 4: Upsampling and across level combination
(determining read and write addresses according to upsampling rounds)

1: **Define:**
2: $R24 - line one read address pointer
3: $R25 - line two read address pointer
4: $R27 - line one write address pointer
5: $R28 - line two write address pointer
6: $R29 - line three write address pointer
7: **if** $R7 == 1 **then**
8:      $R24 = $R17
9:      $R25 = $R24 + $R31
10:     $R26 = 4096
11:     $R27 = $R26
12:     $R28 = $R27 + 1
13:     $R28 = $R28 + $R10
14:     $R29 = $R27 + $R10
15:     $R29 = $R27 + $R10
16: **else if** $R7 == 2 **then**
17:     $R24 = $R18
18:     $R25 = $R24 + $R31
19:     $R26 = 6144
20:     $R27 = $R26
21:     $R28 = $R27 + 1
22:     $R28 = $R28 + $R10
23:     $R29 = $R27 + $R10
24:     $R29 = $R27 + $R10
25: **else if** $R7 == 3 **then**
26:     $R24 = $R19
27:     $R25 = $R24 + $R31
28:     $R26 = 6656
29:     $R27 = $R26
30:     $R28 = $R27 + 1
31:     $R28 = $R28 + $R10
32:     $R29 = $R27 + $R10
33:     $R29 = $R27 + $R10
34: **else**
35:     $R24 = $R20
36:     $R25 = $R25 + $R31
37:     $R26 = 6784
38:     $R27 = $R26
39:     $R28 = $R27 + $R10
40: **end if**

---

implemented in Algorithm 5.19 to Algorithm 5.22 for computational efficiency. After the upsampling process is completed, the register $R7 is checked to determine whether additional up-sampling iterations are required to achieve the dimension of 16×128 pixels. Once the same level map has been upsampled to the dimension of the image strip, the upsampled map is added to the result in S2 (also 16×128 in dimension) as an across level combination described by Algorithm 5.23. The upsampled map for $l = 1$ is the first level map to be added to S2, and therefore,

---

Algorithm 5.13 Stage 4: Upsampling and across level combination
(nearest neighbour interpolation)

1: **while** $R12! = $R31 **do**
2:       $R23 = STRIP_RAM[$R24]
3:       $R24 = $R24 + +
4:       STRIP_RAM[$R27] = $R23
5:       $R27 = $R27++
6:       STRIP_RAM[$R27] = $R23
7:       $R27 = $R27++
8:       STRIP_RAM[$R28] = $R23
9:       $R28 = $R28++
10:      STRIP_RAM[$R28] = $R23
11:      $R28 = $R28++
12:      $R12 = $R12++
13: **end while**
14: $R12 = 0

Algorithm 5.14 Stage 4: Upsampling and across level combination
(four neighbour averaging)

1: **Fill up memory in a checkerboard fashion visualised in 2D**
2: $R22 = $R30 − 1
3: **while** $R11 < $R22 **do**
4:       **do** fill in left border for current row and current row + 2 (Algorithm 5.15)
5:       **do** fill in values from left to right (Algorithm 5.16)
6:       $R27 = $R27 + $R10
7:       $R28 = $R28 + $R10
8:       $R28 = $R28 + 2
9:       $R29 = $R29 + $R10
10:      $R12 = 0
11:      $R11 = $R11++
12: **end while**
13:   **do** fill in most bottom row (Algorithm 5.17)
14:   **Fill up remaining pixels visualised in 2D**
15:   **do** fill remaining pixels (Algorithm 5.18)

Algorithm 5.15 Stage 4: Upsampling and across level combination
(fill left border)

1: $R13 = STRIP_RAM[$R24]
2: $R24 = $R24++
3: $R14 = STRIP_RAM[$R25]
4: $R25 = $R25++
5: STRIP_RAM[$R27] = $R13
6: $R27 = $R27 + 2
7: STRIP_RAM[$R29] = $R14
8: $R29 = $R29 + 2

S2 has to cleared to zero in Stage 3. At the end of Stage 4, all four-level maps of $l = 1$ to $l = 4$ are combined into a single channel map for the luminance feature in S2.

Algorithm 5.16 Stage 4: Upsampling and across level combination
(fill left to right)

```
 1: while $R12 < $R31 do
 2:    if $R12 == ($R31 − 1) then
 3:       $R23 = $R13 + $R14
 4:       $R23 = $R23 >> 1
 5:       STRIP_RAM[$R28] = $R23
 6:    else
 7:       $R15 = STRIP_RAM[$R24]
 8:       $R24 = $R24++
 9:       $R16 = STRIP_RAM[$R25]
10:       $R25 = $R25++
11:       $R23 = $R13 + $R14
12:       $R23 = $R23 + $R15
13:       $R23 = $R23 + $R16
14:       $R23 = $R23 >> 1
15:       $R23 = $R23 >> 1
16:       STRIP_RAM[$R27] = $R15
17:       $R27 = $R27 + 2
18:       STRIP_RAM[$R28] = $R23
19:       $R28 = $R28 + 2
20:       STRIP_RAM[$R29] = $R16
21:       $R29 = $R29 + 2
22:       $R13 = $R15
23:       $R14 = $R16
24:    end if
25:    $R12 = $R12++
26: end while
```

Algorithm 5.17 Stage 4: Upsampling and across level combination
(fill bottom row)

```
 1: $R12 = 0
 2: $R13 = STRIP_RAM[$R24]
 3: $R24 = $R24++
 4: while $R12 < $R31 do
 5:    if $R12 == ($R31 − 1) then
 6:       STRIP_RAM[$R28] = $R13
 7:       $R28 = $R28 + 2
 8:    else
 9:       $R15 = STRIP_RAM[$R24]
10:       $R24 = $R24++
11:       $R23 = $R13 + $R15
12:       $R23 = $R23 >> 1
13:       STRIP_RAM[$R28] = $R23
14:       $R28 = $R28 + 2
15:       $R13 = $R15
16:    end if
17:    $R12 = $R12++
18: end while
```

Algorithm 5.18 Stage 4: Upsampling and across level combination (fill remaining pixels)

1: **Fill up remaining pixels (odd columns)**
2: $R11 = 0
3: $R12 = 0
4: $R28 = $R26 + 1
5: **while** $R11 < $R30 **do**
6:    **while** $R12 < R31 **do**
7:       **if** $R11 == 0 **then**
8:          **do** fill do fill odd column 1 (Algorithm 5.19)
9:       **else**
10:          **do** fill do fill odd column 2 (Algorithm 5.20)
11:       **end if**
12:       $R12 = $R12++
13:    **end while**
14:    $R11 = $R11++
15:    $R12 = 0
16:    $R28 = $R28 + $R10
17: **end while**
18: **Fill up remaining pixels (even columns)**
19: $R11 = 0
20: $R12 = 0
21: $R28 = $R26 + $R10
22: **while** $R11 < $R30 **do**
23:    **while** $R12 < R31 **do**
24:       **if** $R11 == ($R30 − 1) **then**
25:          **do** fill do fill even column 1 (Algorithm 5.21)
26:       **else**
27:          **do** fill do fill even column 2 (Algorithm 5.22)
28:       **end if**
29:       $R12 = $R12++
30:    **end while**
31:    $R11 = $R11++
32:    $R12 = 0
33:    $R28 = $R28 + $R10
34: **end while**

5.4.5   *Stage 5: Salient Pixel Computation*

Stage 5 is the final stage of the VA algorithm. In this stage, the across level combined pixel values in S2 are loaded out, right bit shifted by two to give the salient pixels and stored back into the same locations in S2. The averaged pixels in S2 can then be rearranged back into the 2D format, giving the saliency strip of the input image strip or passed on to an additional stage to be used in a VA-related task. Algorithm 5.24 shows the algorithm of the salient pixel computation process.

Algorithm 5.19 Stage 4: Up-sampling and across level combination
(fill odd column—top border)

```
 1:  if $R12 == ($R31 − 1) then
 2:     $R29 = $R28 − 1
 3:     $R13 = STRIP_RAM[$R29]
 4:     $R29 = $R28 + $R10
 5:     $R14 = STRIP_RAM[$R29]
 6:     $R23 = $R13 + $R14
 7:     $R23 = $R23 >> 1
 8:     STRIP_RAM[$R28] = $R23
 9:     $R28 = $R28 + 2
10:  else
11:     $R29 = $R28 − 1
12:     $R13 = STRIP_RAM[$R29]
13:     $R29 = $R29 + 2
14:     $R14 = STRIP_RAM[$R29]
15:     $R29 = $R28 + $R10
16:     $R15 = STRIP_RAM[$R29]
17:     $R15 = $R15 << 1
18:     $R23 = $R13 + $R14
19:     $R23 = $R23 + $R15
20:     $R23 = $R23 >> 1
21:     $R23 = $R23 >> 1
22:     STRIP_RAM[$R28] = $R23
23:     $R28 = $R28 + 2
24:  end if
```

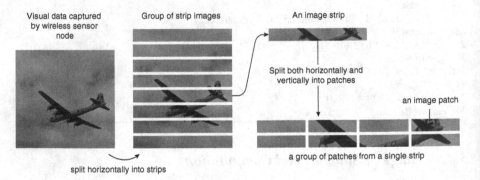

Fig. 5.13 Obtaining image patches from the captured image

5.4.6   Stage 6: Information Reduction Using Salient Patches

The information reduction stage uses the saliency strip of the Stage 5 output to compute the salient patches. The saliency strip provides prediction of salience on a strip basis. Each strip can be further partitioned into smaller sections termed as image patches. Figure 5.13 shows a case where each strip has been further partitioned into eight patches.

Algorithm 5.20 Stage 4: Upsampling and across level combination
(fill odd column—except top border)

```
 1: if $R12 == ($R31 − 1) then
 2:     $R29 = $R28 − 1
 3:     $R13 = STRIP_RAM[$R29]
 4:     $R13 = $R13 << 1
 5:     $R29 = $R28 + $R10
 6:     $R14 = STRIP_RAM[$R29]
 7:     $R29 = $R28 − $R10
 8:     $R15 = STRIP_RAM[$R29]
 9:     $R23 = $R13 + $R14
10:     $R23 = $R23 + $R15
11:     $R23 = $R23 >> 1
12:     $R23 = $R23 >> 1
13:     STRIP_RAM[$R28] = $R23
14:     $R28 = $R28 + 2
15: else
16:     $R29 = $R28 − 1
17:     $R13 = STRIP_RAM[$R29]
18:     $R29 = $R29 + 2
19:     $R14 = STRIP_RAM[$R29]
20:     $R29 = $R28 + $R10
21:     $R15 = STRIP_RAM[$R29]
22:     $R29 = $R28 − $R10
23:     $R16 = STRIP_RAM[$R29]
24:     $R23 = $R13 + $R14
25:     $R23 = $R23 + $R15
26:     $R23 = $R23 + $R16
27:     $R23 = $R23 >> 1
28:     $R23 = $R23 >> 1
29:     STRIP_RAM[$R28] = $R23
30: end if
```

The next step involves salient patch thresholding by comparing the pixel values of the patches with a threshold value given by Eq. (5.28):

$$
\begin{aligned}
&\text{if} \quad P_N(x) > \frac{1}{E}\sum_{p=E}^{p=1} SS(p) \\
&\qquad P_N(x) = 1 \\
&\text{elseif } P_N(x) \le \frac{1}{E}\sum_{p=E}^{p=1} SS(p) \\
&\qquad P_N(x) = 0
\end{aligned}
\tag{5.28}
$$

where $P_N(x)$ is the pixel value at location x in the N-th patch P of the saliency strip SS, $SS(p)$ is the pixel value at location p in SS and E is the total pixels in SS. Each patch is computed for salience by examining the amount of non-zero pixels contained in that particular patch. If the number of non-zero pixels in a patch

---

Algorithm 5.21 Stage 4: Upsampling and across level combination
(fill even column bottom border)

```
 1: if $R12 == ($R31 − 1) then
 2:     $R29 = $R28 − 1
 3:     $R13 = STRIP_RAM[$R29]
 4:     $R29 = $R28 + $R10
 5:     $R14 = STRIP_RAM[$R29]
 6:     $R23 = $R13 + $R14
 7:     $R23 = $R23 >> 1
 8:     STRIP_RAM[$R28] = $R23
 9: else
10:     $R29 = $R28 − 1
11:     $R13 = STRIP_RAM[$R29]
12:     $R29 = $R29 + 2
13:     $R14 = STRIP_RAM[$R29]
14:     $R29 = $R28 − $R10
15:     $R15 = STRIP_RAM[$R29]
16:     $R15 = $R15 << 1
17:     $R23 = $R13 + $R14
18:     $R23 = $R23 + $R15
19:     $R23 = $R23 >> 1
20:     $R23 = $R23 >> 1
21:     STRIP_RAM[$R28] = $R23
22:     $R28 = $R28 + 2
23: end if
```

---

exceeds the minimum number of required pixels, the patch is labelled as salient, else it is labelled as non-salient. This labelling is performed by assigning a one-bit flag called a saliency bit to each of the patches. The patches that have a saliency bit of one associated with them are called salient patches. In the implementation, the number of non-zero pixels in a patch is required to be more than 50 % of the total pixels making up the patch in order for the patch to be classified as salient. At the end of the computation and selection process, the saliency bit for each of the patches from a single saliency strip will be checked iteratively. If a salient patch is detected (saliency bit equals to one), then the saliency bit is transmitted followed by the image patch of the captured image corresponding to the pixel locations defined by the salient patch. If a patch is found to be not salient (saliency bit equals to zero), only the saliency bit is transmitted. When the base station receives a zero-value saliency bit, pixel locations in the reconstructed image corresponding to the strip number and patch will be filled with zeros. Otherwise, the pixel locations will be filled with the data from the received image patch of the captured image. Information reduction is obtained by not transmitting image patches which are found to be non-salient. Image patches which are found to be salient are transmitted in full resolution. Figure 5.14 shows an overview of the information reduction method using salient patches.

At the end of Stage 5, the saliency strip values are located in partition S2. There are three steps in the data reduction computation: computing the mean saliency

---

Algorithm 5.22 Stage 4: Upsampling and across level combination
(fill even column—except bottom border)

```
 1: if $R12 == 0 then
 2:      $R29 = $R28 − 1
 3:      $R13 = STRIP_RAM[$R29]
 4:      $R13 = $R13 << 1
 5:      $R29 = $R28 + $R10
 6:      $R14 = STRIP_RAM[$R29]
 7:      $R29 = $R28 − $R10
 8:      $R15 = STRIP_RAM[$R29]
 9:      $R23 = $R13 + $R14
10:      $R23 = $R23 + $R15
11:      $R23 = $R23 >> 1
12:      $R23 = $R23 >> 1
13:      STRIP_RAM[$R28] = $R23
14:      $R28 = $R28 + 2
15: else
16:      $R29 = $R28 − 1
17:      $R13 = STRIP_RAM[$R29]
18:      $R29 = $R29 + 2
19:      $R14 = STRIP_RAM[$R29]
20:      $R29 = $R28 + $R10
21:      $R15 = STRIP_RAM[$R29]
22:      $R29 = $R28 − $R10
23:      $R16 = STRIP_RAM[$R29]
24:      $R23 = $R13 + $R14
25:      $R23 = $R23 + $R15
26:      $R23 = $R23 + $R16
27:      $R23 = $R23 >> 1
28:      $R23 = $R23 >> 1
29:      STRIP_RAM[$R28] = $R23
30: end if
```

Algorithm 5.23 Stage 4: Up-sampling and across level combination
(across level combination)

```
 1: Define:
 2: $R21 - pixel counter
 3: $R24 - read/write address pointer (reads and writes to S2)
 4: $R25 - read address pointer (reads from S3)
 5: $R26 - number of pixels to be added across level
 6: Initialisation: $R21 = 0; $R24 = 2048; $R25 = 4096; $R26 = 2048
 7: while $R21 < $R26 do
 8:      $R13 = STRIP_RAM[$R24]
 9:      $R14 = STRIP_RAM[$R25]
10:      $R23 = $R13 + $R14
11:      $R24 = $R24++
12:      $R25 = $R25++
13:      $R21 = $R21++
14: end while
```

Algorithm 5.24 Stage 5: Salient pixel computation

1: **Define:**
2: $R3 - read/write address pointer (reads and writes to S2)
3: $R4 - number of pixels to be added to S2
4: $R5 - pixel counter
5: **Initialisation:** $R3 = 2048$; $R4 = 2048$; $R5 = 0$
6: **while** $R5 < $R4 **do**
7:     $R13 = STRIP_RAM[$R3]
8:     $R13 = $R13 >> 1
9:     $R13 = $R13 >> 1
10:     STRIP_RAM[$R13] = $R3
11:     $R3 = $R3++
12:     $R5 = $R5++
13: **end while**

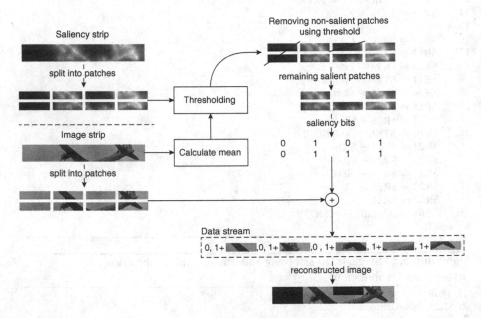

Fig. 5.14 Overview of information reduction using salient patches

value of the saliency strip, pixel thresholding and determining saliency bit. The
operations of the first step are shown in Algorithm 5.25, Lines 8–20. Each location
is cycled through, summing the salience value to a saliency sum register $R7.
After all the locations in the saliency strip have been visited, the value in $R7 is
bit shifted right nine times iteratively to divide the total saliency sum of the strip
by 512 and giving the mean of the saliency strip. In the thresholding step, the
pixels of the saliency strip are thresholded by comparing individual pixel value with
the computed mean saliency value. The pixel values in the saliency strip are first
summed and then divided by 512 to avoid a mean value that is below a value of one.

Algorithm 5.25 Stage 6: Salient patch computation

1: **Define:**
2: $R3 - read/write address pointer (reads froms/writes to S2)
3: $R4 - write address pointer (writes to S3)
4: $R5 - number of pixels to be summed
5: $R6 - pixel counter
6: $R7 - sum of pixels
7: **Initialisation:** $R3 = 2048; $R4 = 4096; $R5 = 2048; $R6 = 0; $R7 = 0
8: **STEP 1: Compute mean salience**
9: $R23 = $R3
10: **while** $R6 < $R5 **do**
11:    $R13 = STRIP_RAM[$R23]
12:    $R7 = $R7 + $R13
13:    $R23 = $R23++
14:    $R6 = $R6++
15: **end while**
16: $R24 = 0
17: **while** $R24 < 9 **do**
18:    $R7 = $R7 >> 1
19:    $R24 = $R24++
20: **end while**
21: **STEP 2: Salient pixel thresholding**
22:    **do** thresholding (Algorithm 5.26)
23: **STEP 3: Determine saliency bit**
24:    **do** determine saliency bit (Algorithm 5.27)

When the condition $R7 == 0 is encountered, the salience value computed in Step 1 is a zero indicating the patch has very few to no salient regions. Therefore, a high value of 64 is used to replace the mean salience value of zero. From observations, patches of dimensions 8×8 consisting of salient regions have a mean salience value between the range of 7–15. The mean salience value can sometimes be found to exceed the value of 32. In such cases, the high values are contributed by small clusters of pixels that centre among neighbouring pixels of diminishing salience values. Often, the high mean salience values for these cases cause the surrounding important pixels of lower values to be thresholded to zero. As a result, a salient patch which should be transmitted falls short of the number of "1"s to qualify for transmission. To address this issue, the salience mean is right bit shifted once when $R7 > 15$. The Step 2 thresholding algorithm is shown in Algorithm 5.26.

Once the threshold has been applied to the salient pixels, they are used to compute the saliency bits. The computation for the saliency bits is found in Algorithm 5.27. Lines 10 and 11 first calculate the start write address for the saliency bits which will be stored in partition S4. In the implementation, there will be 32 patches for each saliency strip. Therefore, the saliency bits will be written to addresses $(8,192 - 32 = 8,160)$ to 8,191. After the start write address for the saliency bits has been computed, each patch is scanned for the total number of "1s" it contains. For a patch that contains full important information, the total number of "1s" will be 64. Algorithm 5.28 cycles through every location in a patch and

Algorithm 5.26 Stage 6: Salient patch computation
(salient pixel thresholding)

1: **Define:**
2: $R8 - threshold value for high average salience values
3: $R9 - new average salience value for condition $R7 == 0
4: **Initialisation:** $R8 = 15; $R9 = 64
5: **Check average salience value**
6: **if** $R7 > $R8 **then**
7:     $R7 = $R7 >> 1
8: **else if** $R7 == 0 **then**
9:     $R7 = $R9
10: **else**
11:     **delay**
12: **end if**
13: **Threshold pixels**
14: $R6 = 0
15: $R23 = $R3
16: $R24 = $R4
17: $R7 = $R7 << 1
18: **while** $R6 < $R5 **do**
19:     $R13 = STRIP_RAM[$R23]
20:     $R13 = $R13 << 1
21:     $R13 = $R13 << 1
22:     **if** $R13 < $R7 **then**
23:         STRIP_RAM[$R24] = 0
24:     **else**
25:         STRIP_RAM[$R24] = 1
26:     **end if**
27:     $R23 = $R23++
28:     $R24 = $R24++
29:     $R6 = $R6++
30: **end while**

---

checks for a "1". For every "1" found, the value of register $R26 is incremented by one. After the checking has been performed, the sum of "1s" in $R26 is returned to Algorithm 5.27. This sum is compared to a predefined transmit threshold value set to 33 and initialised in $R17. If the value in $R26 is larger than the transmit threshold value, then a "1" is written for the saliency bit in S4, else a "0" is written. The address pointing to the next patch is updated when the saliency bit value has been written. Lines 32–36 describe the address updating. After the address of the next patch is determined, the checking for "1s" in a patch and saliency bit value computation are performed again until all patches have been processed.

5.5   Hardware Setup and Results

This section describes the implementation of the VA information reduction technique on the WMSN hardware platform. The VA and the data reduction algorithms are implemented in the custom WMSN as binary instructions loaded into the

---

Algorithm 5.27 Stage 6: Salient patch computation
(determine saliency bit)

1: **Define:**
2: $R11 - patch height (PatchHeight)
3: $R12 - patch width (PatchWidth)
4: $R14 - number of patches per row (NumPatchPerRow = 16 / PatchHeight)
5: $R15 - number of patches per column (NumPatchPerCol = 128 / PatchWidth)
6: $R16 - total number of patches in the strip image (NumPatches = (2048 / (PatchHeight $\times$ PatchWidth))
7: $R17 - transmit threshold (how many 1s in a pacth to qualify for transmission)
8: **Initialisation:** $R11 = 8; $R12 = 8; $R14 = 2; $R15 = 16; $R16 = 32; $R17 = 33
9: **Compute start address for saliency bit in S4**
10: $R18 = 8192
11: $R18 = $R18 - $R16
12: **Begin patch processing**
13: $R19 = 0
14: $R20 = 0
15: $R24 = $R4
16: **while** $R19 < $R14 **do**
17:      **while** $R20 < $R15 **do**
18:          **Determine number of 1s in patch**
19:          **do** extract patch and determine number of 1s (Algorithm 5.28)
20:          **Determine saliency bit**
21:          **if** $R26 < $R17 **then**
22:              STRIP_RAM[$R18] = 0
23:          **else**
24:              STRIP_RAM[$R18] = 1
25:          **end if**
26:          $R24 = $R24 + $R12
27:          $R18 = $R18++
28:          $R20 = $R20++
29:      **end while**
30:      $R29 = 0
31:      **Increment address to next correct location**
32:      $R30 = $R11 - 1
33:      **while** $R29 != $R30 **do**
34:          $R24 = $R24 + 128
35:          $R29 = $R29++
36:      **end while**
37:      $R19 = $R19++
38:      $R20 = 0
39: **end while**

---

instruction memory of the strip-based MIPS processor. Figure 5.15 shows the hardware setup for the VA implementation. The WMSN on the RC10 FPGA board is connected with its ZigBee end device powered by a 9 V battery. The main board of the WMSN is powered by a USB connection. On the receiving end, the ZigBee coordinator is attached to a workstation where the data received by it is sent to MATLAB to be reconstructed into a 2D image for viewing. The coordinator and end device wireless modules operate at a baud rate of 9,600 symbols/s. A packet size of

Algorithm 5.28 Stage 6: Salient patch computation
(determine 1s in a patch)

```
 1: Define:
 2: $R26 - sum of 1s of thresholded salient pixels
 3: $R27 - row patch pixel counter
 4: $R28 - column patch pixel counter
 5: Initialisation: $R26 = 0; $R27 = 0; $R28 = 0
 6: Process patch
 7: $R25 = 0
 8: while $R27 < $R11 do
 9:    while $R28 < $R12 do
10:       $R13 = STRIP_RAM[$R25]
11:       if $R13 == 1 then
12:          $R26 = $R26++
13:       end if
14:       $R25 = $R25++
15:       $R28 = $R28++
16:    end while
17:    $R25 = $R25 + 128
18:    $R25 = $R25 - $R12
19:    $R27 = $R27++
20:    $R28 = 0
21: end while
```

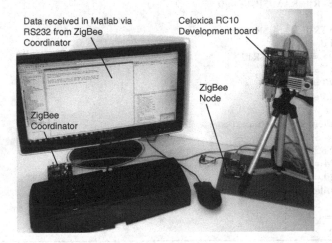

Fig. 5.15 Hardware setup for VA implementation

17 kB and 80 kB is used for the transmission of the saliency bits and the image patch, respectively. For a packet transmitted, 16 kB are occupied by the packet header and checksum.

Figure 5.16 shows some sample of images captured by the WMSN and the outputs after the data reduction. The captured images, saliency maps and the data reduced images are normalised to the range of 0–255 once received in MATLAB.

| Captured image | Strip-based saliency map | Data-reduced image | Data reduction (%) |

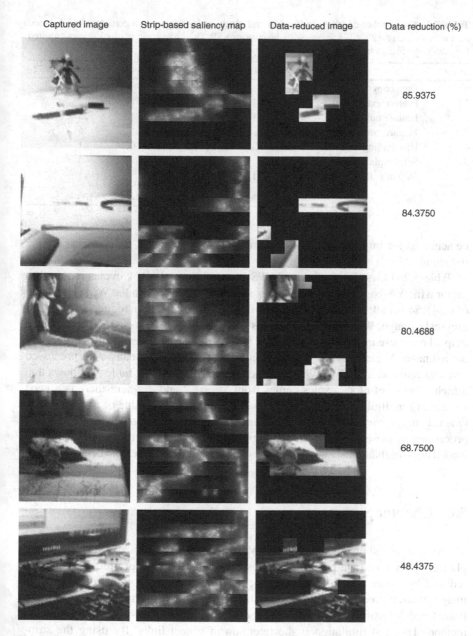

Fig. 5.16 Data reduced images that are captured, processed and transmitted by the WMSN

The normalisation is done to increase the dynamic range of the visual data for viewing clarity as the original pixel values are between the range of 0 and 63 due to the 6-bit image capture. The figure shows that data reduction values up to 85 % can

Table 5.2 The number of MIPS instruction and clock cycles required to perform the VA-based data reduction on a 128×128 image with the custom VSP. The proportion of clock cycles required for each stage is also tabulated

Stage		Instructions	Clock cyles	Proportion (%)
0	Data copy	14	180,296	3
1	Feature extraction	191	753,016	12
2	Feature emphasis	174	1,578,384	25
3	Feature combination	95	300,568	5
4	Upsampling and combination	354	2,524,480	41
5	Salient pixel computation	14	196,664	3
6	Salient patch computation	120	708,136	11
	Total	962	6,241,544	100

be achieved for images with few salient regions. Most images gave at least a 50 % reduction.

Table 5.2 shows the number of MIPS instructions and clock cycles required to perform the VA-based data reduction on a 128×128 image with the WMSN. A total of 6,241,544 clock cycles were required to VA process and data reduce a 128×128 greyscale image. 962 MIPS instructions were required to process a single image strip where these instructions are reused for another seven times to fully process an entire image. At a clock rate of 37 MHz, an image of 128×128 can be VA processed and data reduced in 0.17 s giving a frame rate of 5.8 frames/s. Table 5.2 shows that a high number of clock cycles comes from Stages 2 and 4 where there is a need for loading multiple neighbouring pixels for computing the value of a single pixel. One way to improve this is to use a multiple memory bank approach to speed up the processing of these stages. With this approach, four neighbouring pixels would be loaded out simultaneously along with the centre pixel when given the centre address.

5.6   Chapter Summary

This chapter has shown the potential of VA event processing on a WMSN hardware platform to reduce data prior to transmission. The high-memory requirement was reduced by using a strip-based approach. While many important regions in the images were correctly processed to qualify for transmission, there are two limitations faced by using the strip-based processing approach with the data reduction method. The first limitation is the retention of object links. By using the strip-based approach, each strip is processed individually, thus eliminating the connection of the partial objects within a strip with the other strips. As a result, the partial objects are disjointed from their similar components and cause noticeable isolated boundaries between strips as seen in the saliency maps. Therefore, objects which span across several strips may not qualify for transmission due to the lost of object links. In such images, large regions taking up a significant portion of the strip lose their salience when seen as a strip, but when processed as a whole image, these regions are of high salience values. The second limitation is the false detection arising from illumination, noise, shadows and texture. While these details do not pose any problem to the non-strip-based approach, they are evident in saliency maps computed using the strip-based processing approach. When the saliency map is computed as a whole, important regions and objects are usually large enough to render illumination, noise and fine texture insensitive to salience predictions. However, when the strip-based approach is used, these details are comparable in size to the partial objects contained in the strip causing the locations of these details to have a low saliency value. The technique can be extended for salience predictions in videos to exploit the temporal variations between frames to boost the salience of moving conspicuity objects.

Chapter 6
Single-View Information Reduction Techniques for WMSN Using Event Compression

Abstract This chapter presents single-view information reduction techniques using the set partitioning in hierarchical tree (SPIHT) image compression algorithm. An overview of image compression models using the first and second generation compression algorithms is first described followed by the description of the SPIHT algorithm. Modifications have been introduced on the traditional SPIHT image coding technique with the aim to provide a low-memory implementation of the SPIHT coder in a wireless multimedia sensor network (WMSN) as well as improving its compression performance. The proposed approach employs a strip-based processing technique where an image is partitioned into strips and each strip is encoded separately. Besides, it also uses the new one-dimensional memory-addressing method to store the wavelet coefficients at predetermined locations in the strip buffer for ease of coding. To further reduce the memory requirements, the proposed SPIHT coding employs a new spatial orientation tree (SOT) structure and a listless approach that allow for a very low-memory implementation of the strip-based image coding. In addition, a modification to the SPIHT algorithm by reintroducing the degree-0 zerotree coding methodology was used to give a high-compression performance. Simulations show that even though the proposed image compression architecture using strip-based processing requires a much less complex hardware implementation and its efficient memory organisation uses a lesser amount of embedded memory for processing and buffering, it can still achieve a very good compression performance. The chapter concludes with the hardware implementation of the modified SPIHT coder for low-memory implementation on the strip-based MIPS processor.

6.1 Introduction

A wireless multimedia sensor network (WMSN) consists of many sensing devices that are able to communicate over a wireless channel with the capability of performing data processing and computation at the visual sensor nodes. Multimedia data such as images and videos collected from the visual sensor nodes requires extensive processing and this makes the implementation of a WMSN difficult, especially in a hardware constrained environment. For an example, multimedia data especially high resolution images require extensive bandwidth for transmission. Due to the limited bandwidth available, an image captured by a sensor node needs to be processed and compressed before it is transmitted. The event compression process is carried out to provide a more efficient representation of the image for storage and transmission purposes. With information reduction techniques, the visual data can be transmitted with a higher degree of efficiency. For visual compression in a WMSN, it is essential to maintain a high compression performance while at the same time, providing a low-memory and low-complexity implementation of the visual encoder. The desirable characteristics for the compression algorithm for a WMSN operating in a hardware constrained environment include:

- *Efficient compression performance.* The proposed algorithm should be able to give a good peak signal-to-noise ratio (PSNR) performance that is comparable to existing popular image and video coding algorithms.
- *Signal-to-noise ratio (SNR) scalability.* SNR scalability is a feature in the encoded bitstream that allows the decoder to decode an image to different levels of quality. The more the bitstream is being decoded, the higher will be the reconstructed image quality.
- *Lossy to lossless coding.* The proposed algorithm should be able to provide a lossy up to a lossless coding in a single encoded bitstream.
- *Low-memory requirement.* The proposed work should allow coding to be carried out on only a portion of an image at a time to achieve low-memory implementation. The individual bitstreams obtained from each portion of the image are then grouped together to form a final embedded bitstream prior to transmission.
- *Low complexity.* For implementation in a hardware constrained environment, the complexity of the proposed algorithm should be as low as possible. Besides, a coder with a lower complexity helps to minimise the implementation cost and requires less computations.
- *Hardware implementable.* In this chapter, the hardware architecture of a modified SPIHT algorithm will be implemented using the microprocessor-based approach on a field programmable gate array (FPGA) device.

The image or video data to be compressed can be obtained from a single camera or from multiple cameras of overlapping scenes. The single-view setup involves an image capturing device which is set at a predetermined location and angle to capture a scene. In general, single-view information reduction techniques using image coding methods involve a few processing stages as shown in Fig. 6.1. An image captured by the camera is first transformed to a domain that is more suitable for processing. Quantization is then carried out on the transformed image data. Next, the image compression algorithm is applied followed by entropy coding. This generates a bitstream that is ready for transmission. At the decoder side, the above processes are reversed which finally leads to the reconstruction of the original image.

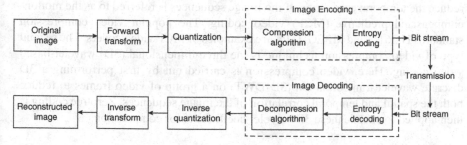

Fig. 6.1 Overview of single-view event compression process

6.2   Visual Compression and Models

This section will briefly review some concepts in visual compression such as spatial and temporal redundancy, lossless and lossy compression and image quality. The section will also briefly review image compression models and algorithms which have been proposed in the literature.

6.2.1   Spatial and Temporal Redundancy

Visual compression involves the process of discarding the redundant information by finding a less correlated representation for a single-image or for a sequence of video data. The compression techniques revolve around two basic concepts which are spatial redundancy and temporal redundancy. In spatial compression, information reduction is performed on each individual frame of an image sequence. This means that compression is applied on a single image as a completely independent entity with no relation to other frames. This coding process that is applied on a two-dimensional (2D) image data is also referred to as intraframe compression. On the other hand, temporal compression happens over a series of image frames and takes advantage of areas of the image that remain unchanged from frame to frame, and discarding the data for repeated pixels. The temporal reduction technique focuses on storing the changes between subsequent frames rather than the entire frame in isolation. Videos without a lot of motion, such as talking head clips, take the best advantage of temporal compression which is also referred to as interframe compression.

For image coding, it is straight forward where only spatial compression is performed on the single-image frame. However, in video coding, the compression techniques can be divided into two types. The first type involves the process of motion estimation to remove the redundant image data between the video frames. During motion estimation, the movement of objects in an image sequence is first analysed and the predicted motion vector is then encoded to achieve compression. This type of video coding technique that applies a motion estimation technique to reduce the temporal correlation of the image sequences is referred to as the motion-compensated predictive (MCP) video coding. The popular video compression standards H.261/3 and MPEG-1/2/4 are examples of MCP video coders. The second type of video compression technique is the three-dimensional (3D) wavelet-based video coding. Here, video compression is carried out by first performing a 3D discrete wavelet transformation (3D-DWT) on a group of video frames to reduce both the spatial and temporal correlation of the image sequences. Zerotree coding is then applied to encode these 3D wavelet coefficients for compression.

6.2.2 Lossy and Lossless Compression

Compression can be lossy or lossless. Lossy compression sacrifices some data from the file to achieve a much higher compression rate. In image processing applications such as web browsing, photography, image editing and printing, a lossy coding is sufficient as an image compression tool. Although some information loss can be tolerated in most of these applications, there are image processing applications that require no pixel differences between the original and the reconstructed image. Such applications include medical imaging, remote sensing, satellite imaging and forensic analysis where a lossless form of compression is very important. In lossless compression, no information is lost and the final file is identical to the original.

6.2.3 Evaluation of Image Quality

Image quality is a characteristic of an image that measures the perceived image degradation compared to an ideal or perfect image. Two different ways of assessing image quality are the subjective approach and the objective approach. Subjective measurement involves a human observer where the image quality is assessed by asking human subjects for their opinions. This type of image quality measurement perceived by human observers is becoming increasingly important due to the large number of applications that target humans as the end users. In objective measurement, the image quality is assessed by means of the peak signal-to-noise ratio (PSNR). The PSNR is a function of the mean squared error (MSE) between the original and the reconstructed images. In this chapter, the image compression performance of the proposed work is evaluated in terms of the PSNR as given in Eq. (6.1) where the MSE is calculated using Eq. (6.2).

$$PSNR_{greyscale}(dB) = 10 \log_{10} \frac{255^2}{MSE} \tag{6.1}$$

$$MSE = \frac{(Image_{original} - Image_{reconstructed})^2}{Row_{image} \times Column_{image}} \tag{6.2}$$

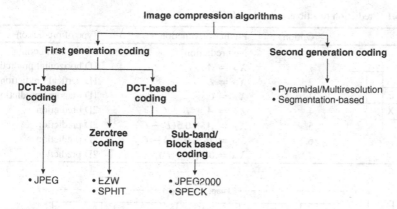

Fig. 6.2 Types of image compression algorithms

6.2.4   Image Compression Models

Image compression models can be categorised under first generation and second generation image coding. First generation image coding emphasises more on how well the information contained in a transformed image is efficiently encoded, whereas the second generation image coding places more importance on how we can exploit and extract useful information from the image. The second generation image coding makes use of available techniques developed in the image compression stage to encode the sequence of information obtained from the image transformation stage. Some of these popular image compression schemes are summarised in Fig. 6.2. Examples of first generation image coding schemes are those based on the discrete cosine transform (DCT) and discrete wavelet transform (DWT). DWT-based coding can be further subdivided into zerotree coding methods or sub-band/block-based coding methods. Examples of second generation coding schemes are pyramidal/multi-resolution and segmentation-based coding schemes. In the next section, we will briefly review several image compression models.

6.2.4.1   Joint Photographic Experts Group Image Coding (JPEG)

The Joint Photographic Experts Group (JPEG) [132] uses the DCT based image compression technique. Since JPEG became an international standard in 1992, it has been widely used in many imaging applications such as internet browsing, photography, image editing and printing. JPEG defines four modes of operation: sequential predictive mode, sequential DCT-based mode or baseline JPEG, progressive DCT-based mode and hierarchical mode. Among the four, the sequential predictive mode provides lossless compression, whereas the other three modes provide lossy compression. In lossless JPEG coding, the current pixel value (X)

Table 6.1 Prediction functions in lossless JPEG

	Option	Prediction function	Type of prediction
	0	No prediction	Differential coding
	1	$X_P = A$	1D horizontal prediction
	2	$X_P = B$	1D vertical prediction
	3	$X_P = C$	1D diagonal prediction
	4	$X_P = A + B - C$	2D prediction
	5	$X_P = A + 0.5(B - C)$	2D prediction
	6	$X_P = B + 0.5(A - C)$	2D prediction
	7	$X_P = 0.5(A + B)$	2D prediction

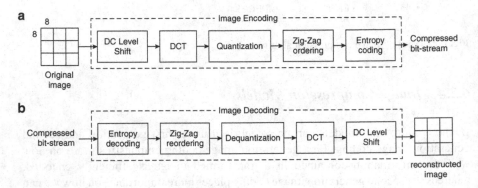

Fig. 6.3 Baseline JPEG. (**a**) Compression. (**b**) Decompression

is predicted based on the previously coded pixel values in its horizontal (A), vertical (B) and diagonal (C) directions. Table 6.1 shows the prediction functions used. After X is predicted, the difference between the predicted value and the original value which is referred to as the prediction error is then entropy encoded using a variable-length encoding technique. The compression quality is dependent on the nature of the image and the type of predictors used.

Figure 6.3 shows the block diagram for baseline JPEG coding. The image is first divided into blocks of 8 × 8 pixels. Each of the 64 pixels is subtracted by 128 to convert the pixel value into a signed integer. The DCT is then applied on each of the image blocks to decorrelate the image pixels. During the quantization stage, information loss is introduced. Following the quantization process, a zigzag scanning is applied to rearrange the coefficients from the lower frequency components to the higher frequency components prior to entropy coding. The low-frequency component is then encoded by differential coding and the higher frequency components are Huffman encoded in the form of (RUNLENGTH, CATEGORY) (AMPLITUDE), where RUNLENGTH is the number of consecutive zeros preceding a non-zero AC coefficient, CATEGORY is the number of bits that represents the variable-length integer code of this non-zero AC coefficient and AMPLITUDE is the unsigned value of the non-zero AC coefficient. The baseline JPEG decompression is the reverse of the compression process.

Unlike the sequential coder, the JPEG progressive coder carries out scanning of an image multiple times. In each scan, a finer detailed version of the image information is transmitted. The DCT is first performed on every block of the whole image and each block contains 8×8 pixels. The transformed image is then stored into a memory buffer for multiple-pass encoding. There are two types of progressive coding modes: the spectral selection and the successive approximation modes. In the spectral selection mode, the DCT coefficients are encoded starting from the lower frequency components and moving progressively to the higher frequency components. In the successive approximation mode, a certain number of the most significant bits of all the DCT coefficients of all the blocks are first encoded and transmitted in the first scan, followed by the remaining bits of all the DCT coefficients of all the blocks in the subsequent scans. JPEG coding in hierarchical mode has the property of resolution scalability. In this mode, the original image is first filtered and down-sampled to the target resolution. For example, the original image which has a size of $N \times N$ pixels is first down-sampled to a size of $N/2 \times N/2$ pixels. This lower resolution image is then encoded using any of the other three JPEG modes. Following this, the $N/2 \times N/2$ image is decoded and up-sampled. This up-sampled decoded image is then compared to the image in the higher resolution. The difference between these two images is then encoded using any of the other three JPEG modes.

6.2.4.2 Embedded Zerotree Wavelet-Based Image Coding (EZW)

Embedded coding began to gain attention in the field of image coding with the intro-duction of the wavelet-based image coding technique called the Embedded Zerotree Wavelet (EZW) algorithm [113]. EZW coding first involves the discrete wavelet transformation (DWT) process which is performed on an image to decompose it into multi-resolution wavelet sub-bands that contain the high- and low-frequency components of the image. In the wavelet sub-bands, every coefficient at a given scale is related to a set of coefficients at the next lower scale of similar orientation. The coefficient in the higher scale is known as the parent and the coefficients in the next lower scale in the same orientation are called the children or offspring. All children and grandchildren under a parent node are referred to as the descendants of the parent node. Figure 6.4a shows the definition of coefficients in a tree.

Figure 6.4b shows the parent–children relationship of the wavelet coefficients in the higher scale and lower scale. Each node will have 2×2 children at its lower scale of similar orientation corresponding to the same location except for the lowest scale where the coefficient does not have any descendant. At the highest scale, every node in the low–low (LL) sub-band will only have three children, one in each of the highest high–low (HL), low–high (LH) and high–high (HH) sub-bands. Due to the nature of the wavelet decomposition, the higher scale sub-bands will contain more energy than the lower scale sub-bands. Thus, the embedded coding starts with the highest LL sub-band followed by HL, LH and HH sub-bands. Figure 6.4b shows the scanning order where the sub-bands in the $(N - 1)$ level are scanned only after

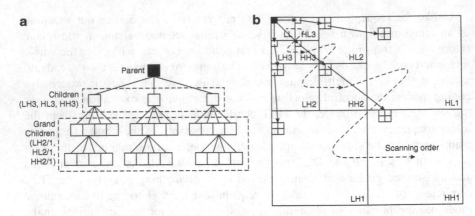

Fig. 6.4 Tree structure and parent–children dependencies of sub-bands in EZW coding. (a) Definition of coefficients in a tree. (b) Zigzag scanning

scanning in the N level is completed. The EZW coding involves the coding of the position of those wavelet coefficients that will be transmitted as a non-zero value. A wavelet coefficient, x is said to be insignificant with respect to a given threshold T if $|x| < T$. However, if $|x| \geq T$, then the coefficient is said to be significant with respect to T. The algorithm is developed based on the following hypothesis taken from [113]—"if a wavelet coefficient at a higher scale is insignificant with respect to a given threshold T, then all the wavelet coefficients of the same orientation in the same spatial location at the lower scales are likely to be insignificant with respect to T." In EZW coding, a tree is referred to as a zerotree if the parent node and all its descendants are insignificant with respect to the threshold. A coefficient of a zerotree for a threshold T is coded as zerotree root (ZTR) if it is not the descendant of a previously found ZTR for that threshold T. However, if a coefficient is insignificant with respect to threshold T but has some significant descendants, it is then coded as isolated zero (IZ). For a coefficient that is found to be significant with respect to T, it is coded as positive significant (POS) or negative significant (NEG) depending on the sign of the coefficient. Figure 6.5 shows the flow chart for EZW coding.

EZW encoding starts with an initial threshold, T_0 which is normally equal to 2^n and is given in Eq. (6.3) where X_j is the largest coefficient found in the wavelet-transformed image. The encoding is then continued under a sequence of thresholds $T_0, T_1, T_2, T_3 \ldots T_{N-1}$ where T_i is set to half of the previous threshold. This encoding methodology is also referred to as bit-plane encoding.

$$T_0 > \frac{|X_j|}{2} \tag{6.3}$$

6.2.4.3 Set Partitioning in Hierarchical Tree (SPIHT)

Wavelet-based image compression algorithm based on set partitioning in hierarchical trees (SPIHT) [111] was introduced by A. Said and W. A. Pearlman

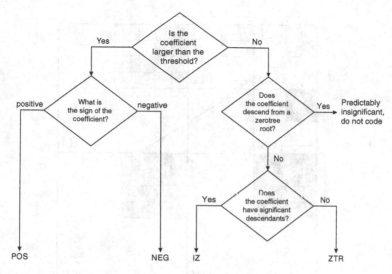

Fig. 6.5 Flow chart for EZW coding

in 1996. This coding scheme operates by exploiting the relationships among the coefficients across the different scales at the same spatial location in the wavelet sub-bands. SPIHT coding adopts a set partitioning approach where the significance test result is binary. A magnitude test is carried out on every wavelet transformed coefficient, C at coordinate (i, j), to indicate the significance of the coefficient. A coefficient is encoded as significant, i.e. $S_n(i, j) = 1$ if its value is larger than or equal to the threshold T, or as insignificant, i.e. $S_n(i, j) = 0$ if its value is smaller than T. This significance test is given in Eq. (6.4) where $(n + 1)$ is the number of bits used to represent the largest coefficient.

$$S_n(i, j) = \begin{cases} 1 & \text{when, } \max\{C_{i,j}\} \geq 2^n \\ 0 & \text{otherwise} \end{cases} \tag{6.4}$$

The parent–children relationship for the SPIHT algorithm is shown in Fig. 6.6. Each node in the wavelet decomposition sub-bands has 2×2 offspring in the same spatial orientation in its lower scale except for the leaf nodes. The tree roots in the highest level are grouped into 2×2 adjacent pixels and in each group, one of them has no descendant.

The sets of coordinates that are defined in SPIHT coding are:

$O(i, j)$	—	holds the set of coordinates of 2×2 offsprings of node (i, j).
$D(i, j)$	—	holds the set of coordinates of all descendants of node (i, j).
$L(i, j)$	—	holds the set of coordinates of all grand descendants of node (i, j), i.e. $L(i, j) = D(i, j) - O(i, j)$.
H	—	holds the set of coordinates of all spatial orientation tree roots.

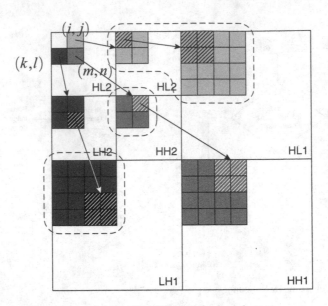

Fig. 6.6 Parent–offspring dependencies of SPIHT spatial orientation tree

The order of subsets which are tested for significance is stored in three ordered lists: The LSP and LIP contain the coordinates of individual pixels, whereas the LIS contains either the set $D(i, j)$ or $L(i, j)$ which is referred to as Type A or Type B, respectively. SPIHT coding first defines a starting threshold according to the maximum value of the wavelet coefficients. The LSP is set as an empty list and all the nodes in the highest sub-band are put into the LIP. The root nodes with descendants are put into the LIS. Two encoding passes which are the sorting pass and the refinement pass are performed in the SPIHT coding. During the sorting pass, a significance test is performed on the coefficients according to the order in which they are stored in the LIP. Elements in LIP that are found to be significant with respect to the threshold are moved to the LSP list. A significance test is then performed on the sets in the LIS. If a Type A set in LIS is found to be significant, the set is removed from the list and is partitioned into four single elements and a Type B set. This Type B set is added back to LIS and the four elements are then tested and moved to LSP or LIP depending on whether they are significant or insignificant with respect to the threshold. However, if a Type B set in LIS is found to be significant, the set is removed and is partitioned into four Type A sets which are then added back to the LIS. The refinement pass is then carried out on every coefficient that is added to the LSP except for those that have just been added during the sorting pass. Each of these coefficients in the list is refined to an additional bit of precision. Then, the threshold is halved and SPIHT coding is repeated until all the wavelet coefficients are coded for lossless compression or until the target rate is met for lossy compression. The SPIHT algorithm reproduced from [111] is shown in Algo 6.1.

Algorithm 6.1 SPHIT algorithm

Step 1. Initialization:
Output $n = [log_2(max(i, j)|Ci, j|)]$
3: LSP is set as an empty list.
Coordinates $(i, j) \in H$ is added to the LIP and those nodes with descendants are added to LIS
as Type A entries.

Step 2. Sorting pass:
6: **for all** entry, (i, j) in the LIP **do**
    Output $S_n(i, j)$
    if $S_n(i, j) = 1$ **then**
9:       move (i, j) to the LSP and output the sign of $C_{i,j}$
    end if
end for

12: **for all** entry, (i, j) in the LIS **do**

    if entry = Type A **then**
       Output $S_n(D_n(i, j))$
15:      **if** $S_n(D_n(i, j)) = 1$ **then**
          for all $(k, l) \in O(i, j)$ **do**
             Output $S_n(k, l)$ and
18:             **if** $S_n(k, l) = 1$ **then**
                add (k, l) to the LSP and output the sign of $C_{k,l}$
             end if
21:             **if** $S_n(k, l) = 0$ **then**
                add (k, l) to the end of the LIP
             end if
24:          **end for**
          if $L(i, j) \neq \varnothing$ **then**
             move $(i.j)$ to the end of the LIS as an entry of Type B, and
27:             go to Step 2.5
          else
             remove entry (i, j) from the LIS
30:          **end if**
        end if
    else if entry = Type B **then**
33:       Output $S_n(L(i, j))$
       if $S_n(L(i, j)) = 1$ **then**
          add each $(k, l) \in O(i, j)$ to the end of the LIS as an entry of Type A, and
36:          remove (i, j) from the LIS
       end if
    end if
39: **end for**

Step 3. Refinement pass:
For each entry (i, j) in the LSP, except those included in the last sorting pass (i.e. with the
same n), output the n-th most significant bit of $|C_{i,j}|$

42: **Step 4. quantization-step update:**
Decrement n by 1 and go to Step 2.

The set partitioning approach that is used by SPIHT is an efficient and compact technique that has a better compression performance than EZW even without arithmetic coding. Since both the encoder and decoder share the same algorithm, i.e. the execution path of the encoder is repeated at the decoder, the computational complexity at the encoder and decoder is the same. SPIHT coding also allows progressive transmission to take place since the wavelet coefficients are transmitted in the order of magnitude with the most significant bit sent first. This generates an embedded bitstream and allows a variable bit rate and rate distortion control. Although the decoder can stop decoding at any point in the bitstream, a fully reconstructed image can still be obtained.

6.2.4.4   Joint Photographic Experts Group 2000 (JPEG 2000)

The JPEG 2000 [114] compression system can be divided into two main stages: the image pre-processing stage and the compression stage. Figure 6.7 shows the JPEG 2000 compression process. The image pre-processing stage is further subdivided into the tiling process, DC level shifting and the multi-component transformation. To reduce the memory requirement in the implementation of the coder, the JPEG 2000 coding divides an input image into rectangular non-overlapping blocks. These blocks are referred to as "tiles" and each of these tiles is then coded independently. Tiling causes blocking artefacts in the reconstructed image which can be reduced by using larger tiles but at the expense of incurring a larger amount of memory buffer. To ease the mathematical calculation at the later part of the compression stage, the unsigned image pixels are DC shifted. By subtracting 2^{P-1} from the actual pixel value where P is the precision of the image pixel, the dynamic range of the original unsigned image is shifted until it is centred at around zero. For colour images, the red–green–blue (RGB) components are transformed into luminance and chrominance components (YUV or YC_bC_r) in order to reduce the correlations among the RGB colour components. JPEG 2000 Part 1 supports the reversible colour transformation (RCT) for lossless compression and irreversible colour transformation (ICT) for lossy compression. By reducing the correlations among the colour components, a higher compression ratio can be achieved since redundancy can be efficiently removed.

During the compression stage, a DWT is first performed on the image followed by the quantization process. The DWT is applied to decompose the image tiles into wavelet sub-bands at different resolution levels. For lossy compression, a CDF 9/7 biorthogonal filter is used and for lossless compression, JPEG 2000 uses the reversible Le Gall 5/3 filter. After the DWT, the wavelet coefficients are quantized to reduce the precision of the sub-bands. The quantization process is only carried out for lossy compression since it will introduce losses. The wavelet sub-bands are then divided into code blocks typically of size 32×32 or 64×64 pixels. Each of these code blocks are then entropy coded independently. Depending on the application needs, JPEG 2000 is able to provide Region of Interest (ROI) coding where the different parts of an image can be coded with a different degree of

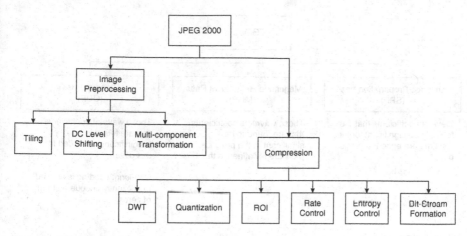

Fig. 6.7 JPEG 2000 compression

fidelity. To do this, the selected region is first shifted to a higher amplitude. This technique is called the MAXSHIFT technique. During entropy coding, the ROI region is encoded prior to other regions since it has a higher magnitude level. At the decoding stage, the selected region is also decoded prior to other regions. During the rate control process, each code block is allocated a predefined bit rate. This is done so that the overall targeted bit rate can be achieved and the image is reconstructed with minimum distortion. Finally, the bitstreams generated from the entropy coder are rearranged into layers and the compressed JPEG 2000 bitstream is transmitted.

JPEG uses the Embedded Block Coding with Optimized Truncation (EBCOT) image compression technique [120]. In EBCOT, the information coding and the information ordering are separated and are referred to as two-tier coding. Tier 1 performs the entropy coding. EBCOT uses the bit-plane coding methodology. Each of the coefficients in the code block is scanned and is encoded in each encoding pass. The process of EBCOT coding is divided into three coding stages as shown in Fig. 6.8: significance propagation pass (SPP), magnitude refinement pass (MRP) and cleanup pass (CUP). The bitstream generated by the EBCOT coder is further encoded by a context-based adaptive binary arithmetic MQ-coder. Tier 2 performs the bitstream formation where the independently generated bitstreams from Tier 1 are subdivided into small chunks. These chunks are then interleaved and packed into different quality layers depending on their contributions to the layers. At the end of the Tier 2 process, a compressed bitstream is obtained and is ready for transmission. Although JPEG 2000 provides many features such as embedded coding, progressive transmission, SNR scalability, resolution scalability, ROI coding and random access features, its multi-layered coding procedures are very complex and require high computations.

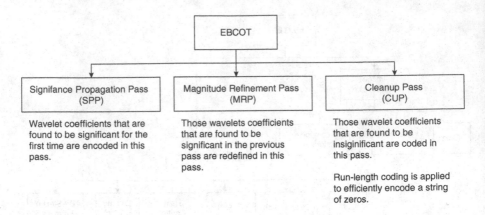

Fig. 6.8 EBCOT coding

6.2.4.5 Set Partitioning Embedded Block Coder (SPECK)

The set-partitioning embedded block coder (SPECK) [101] is a block-based coding scheme that has many similarities with the SPIHT algorithm. This algorithm uses the set partitioning approach where a group of wavelet coefficients are tested and are partitioned into smaller groups during the encoding process. Two lists known as the list of significant pixels (LSP) and list of insignificant sets (LIS) are used to hold the order in which the coefficients and set of coefficients are to be refined and tested, respectively. However, unlike the SPIHT algorithm, the SPECK coding scheme does not exploit the similarity of wavelet coefficients across the different sub-bands of a spatial orientation tree. It codes the wavelet coefficients in each sub-band in a block-by-block basis. The SPECK coding methodology is shown in Fig. 6.9.

Since energy is more likely to be concentrated at the top most levels of the pyramidal structure of wavelet sub-band decomposition, the SPECK coding starts off by partitioning the wavelet-transformed image into two rectangular regions, Type S and Type I, as shown in Fig. 6.9(Step 1). The Type S set is then added into the LIS. The set S in the LIS is subsequently tested for significance against a threshold. If set S is found to be significant, it is further partitioned into four subregions as shown in Fig. 6.9(Step 2). These sub-regions are then inserted back into the LIS for the significance test. However, if set S is found to be insignificant against the threshold, it remains in the LIS and this set is tested again during the next encoding pass. When the set S has been partitioned until it reaches a single pixel, the significant pixel is moved into the LSP and the insignificant pixel is retained in the LIS. After all the Type S sets in LIS have been tested, Type I set is then tested for significance against the same threshold. If Type I set is found to be significant, it is partitioned into four sets consisting of three sets of Type S' and one set of Type I' as shown in Fig. 6.9(Step 3). This partitioning approach is referred to as the octave band partitioning. Similarly, the three sets of Type S' are inserted into the LIS for significance tests. The coding process of Type S and Type I is continued until

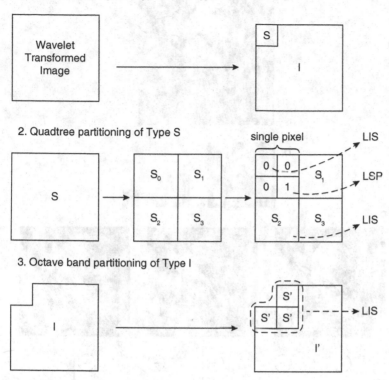

Fig. 6.9 SPECK coding methodology

all the sets have been tested. The refinement process performed on the coefficient in the LSP is similar to the SPIHT algorithm. At the end of each encoding pass, the threshold is halved and the coding process is repeated for those coefficients or sets of coefficients in the two lists. Before transmission, the binary output bits from the SPECK coder are entropy encoded using a context-based arithmetic coding. Since the SPECK coder generates an embedded bitstream, it can provide quality scalability and progressive transmission.

6.2.4.6   Pyramidal/Multi-resolution Coding

Pyramidal coding can be considered as a second generation image coding approach since the hierarchical coding methodology is similar to the nervous system of our human visual system (HVS). Figure 6.10 shows the process of pyramidal coding. The Laplacian pyramid algorithm breaks up an image into components based on the spatial frequency. The value at each node in the pyramid represents the difference between two Gaussian-like functions convolved with the original image. Pyramidal coding provides a multi-resolution representation of an image.

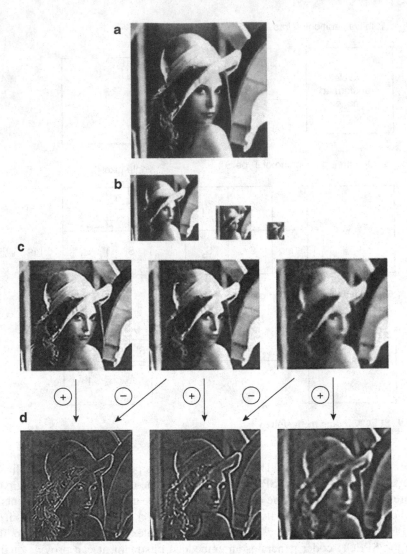

Fig. 6.10 Process of pyramidal coding. (**a**) Original image Lenna. (**b**) Gaussian pyramid image. (**c**) Gaussian interpolation image. (**d**) Laplacian pyramid image

A low-resolution image that contains low-frequency components is first transmitted. This low-resolution image contains most of the energy of the image and it can be encoded with relatively fewer bits. High-frequency components that are encoded at a higher resolution are then progressively transmitted. However, on certain images, these high-frequency features such as points, edges and lines, may require many bits to encode. In this situation, pyramidal coding will become less efficient as an encoding algorithm.

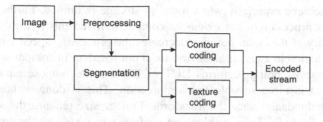

Fig. 6.11 Process for segmentation-based coding

6.2.4.7 Segmentation-Based Coding

By utilising the fact that the human eye is good at identifying regions that are similar and grouping them accordingly, this coding approach works by partitioning the image into sub-regions based on their texture structure. These regions are surrounded by contours and both the contour and texture regions are coded separately. Figure 6.11 shows the process for segmentation-based coding. Pre-processing is first carried out on the image to remove noise as well as the non-useful regions based on the characteristics and analysis of the HVS. During the segmentation process, each pixel and its neighbouring pixels are examined to determine if they share the same properties based on their greyscale levels. Pixels with the same properties are assigned to the same region. This region-growing process is repeated until all the pixels in the image have been assigned to some region. Weakly contrasted adjacent regions and small regions are then merged to reduce the number of regions obtained. Finally, contour coding and texture coding are then applied on the contour and texture regions, respectively. One of the major difficulties faced in the implementation of segmentation-based coding is the setting of thresholds which is used to determine if two pixels have the same properties because the threshold values depend on image content.

6.2.4.8 Comparison of Image Compression Models

Second generation image compression algorithms incorporate the properties of HVS into the coding system to identify the features within the image and then process these features to achieve compression. This characteristic of the second generation image coding that emphasises on exploring the content of an image and makes use of this information to achieve compression requires a more complex and extensive image processing compared to the first generation image coding which performs DCT or DWT. Also, most of the second generation image compression algorithms provide lossy compression and they rely heavily on the initial segmentation [72, 109]. During the segmentation process, image pixels are first classified into contour and texture regions followed by a region-growing process. The whole image is expected to be stored and available in memory during pre-processing and this is difficult to achieve especially when memory storage is limited. Furthermore, the segmentation process requires extensive computation and this not only increases the complexity of the coder but also decreases the processing speed which makes its implementation in a real-time environment not feasible. In comparison, the first generation image coding performs DCT or DWT to decompose an image into a domain that is more suitable for compression. This is done without creating any excess redundancy since the transformed image size remains the same as its original size. Since DCT causes blocking artefacts which reduces the quality of the reconstructed image, the DWT-based image coding is generally preferred over the DCT-based image coding.

Sub-band coding such as EBCOT which is used in the JPEG 2000 standard provides a higher compression efficiency compared to many zerotree coding techniques. However, its multi-layered coding procedures are very complex and computationally intensive. Also, the need for multiple coding tables for arithmetic coding requires extra memory allocation which makes the hardware implementation of the coder more complex and expensive [3]. Although other sub-band coding techniques such as SPECK has a lower coding complexity than EBCOT, arithmetic and Huffman coding are still part of the necessary component in the implementation of sub-band coding systems. EZW and SPIHT are two of the earliest and most popular image compression algorithms that apply the concept of zerotree coding. Both of these algorithms use the bit-plane coding methodology. Their embedded property of being able to generate bitstreams according to the order of importance allows variable bit rate coding and decoding. Thus, an exact targeted bit rate or tolerated distortion degradation can be achieved since the encoder or decoder can stop at any point in the bitstream processing. Furthermore, EZW and SPIHT coders also allow progressive transmission to be carried out. Comparing EZW and SPIHT, the latter not only has all the advantages that EZW possesses, it has also been shown that SPIHT provides a better compression performance than EZW coding. In addition, the SPIHT algorithm has straightforward coding procedures and requires no coding tables since arithmetic coding is not essential in SPIHT coding.

6.3 SPIHT Coding with Improvements

There is a growing interest in improving the performance and efficiency of the original SPIHT algorithm [46, 135]. The improvements can be classified into two categories: (a) improvement on hardware implementation which includes the design of a low cost, low complexity and low-memory image codec and (b) improvement on compression ratio. In this section, several modifications which result in a significant improvement to SPIHT coding in terms of memory requirement and compression efficiency will be discussed.

6.3.1   Low-Memory Implementation of SPIHT Coding

In many wavelet-based image coding applications where an external memory bank
is available, the wavelet-transformed image is first stored into the external memory.
During the image coding process, the required image data is then transferred from
the external memory bank into a temporary memory bank for processing. Although
memory constraint may not be an issue in the implementation of such a coder,
the image data that needs to be transferred for image processing should be as
less frequent as possible. This is because expensive input/output (I/O) is needed
for the data transfer and frequent memory access increases the processing time of
the coder. However, the situation is different when it comes to the implementation
in a WMSN where the image coding needs to be carried out in a memory
constrained environment. When there is a limited amount of external memory
available, a portion of the image data that outputs from the image sensor needs to be
processed as soon as the data required for processing is buffered into the temporary
memory. This portion of image data is then cleared from the memory immediately
after image coding and the next portion of image data is then loaded into the
memory for processing. In this subsection, some modifications are proposed on the
traditional SPIHT coding for memory reduction in its hardware implementation.

6.3.1.1   Strip-Based SPIHT Coding

In the traditional SPIHT image compression, an image stream is first input into the
DWT module. The computed N-decomposition wavelet coefficients that provide a
compact multi-resolution representation of an image are then obtained and stored in
a memory bank. SPIHT coding is subsequently performed to obtain an embedded
bitstream. This processing method would require the whole image to be stored in
a memory for the SPIHT compression. The strip-based coding technique proposed
in [14] which adopts the line-based wavelet transformation proposed in [32] has
shown great improvements in low-memory implementation for SPIHT compression.
In strip-based coding, an image is acquired in raster scan format from an image
sensor and the computed wavelet coefficients are buffered in a strip buffer for the
SPIHT coding. Once the coding for the strip buffer is completed, it is released and
ready to receive the next set of data lines. Since only a portion of the full wavelet
decomposition sub-band is encoded at a time, there is no need to wait for the full
transformation of the image, and coding can be performed once a strip is fully
buffered. This enables the coding to be carried out rapidly and also greatly reduces
the memory storage needed for the strip-based SPIHT coding. Figure 6.12 shows the
strip-based processing where the SPIHT coding is carried out on part of the wavelet
coefficients that are stored in a strip buffer.

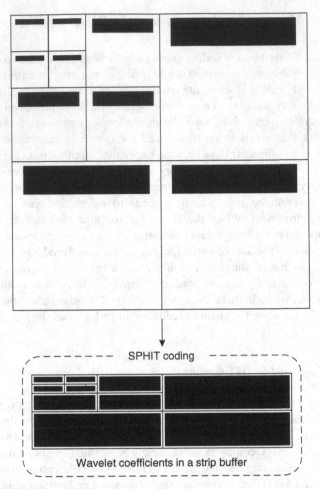

Fig. 6.12 SPIHT coding is carried out on part of the wavelet coefficients that are stored in a strip buffer

6.3.1.2 New Spatial Orientation Tree Structure

As a result of the wavelet transformation, a pyramid structure of wavelet coefficients that represents the spatial relationship between the tree nodes in the higher sub-bands and lower sub-bands is obtained. This tree structure is termed the spatial orientation tree (SOT) structure. In the original SPIHT SOT structure, each node has 2×2 adjacent pixels of the same spatial orientation as its descendants. The parent–children relationship of the original SPIHT SOT is given in Eq. (6.5a). Here, new SOT structures are proposed. For certain sub-bands, the new tree structures take the next four pixels of the same row as their children. The parent–children relationship of the proposed new tree structures is given in Eq. (6.5b). These new tree structures are named SOT-B, SOT-C, SOT-D and so on depending on the number of scales the

Table 6.2 Tree structure

Algorithm	Tree structure
SOT-A	Traditional SPIHT tree structure. Each node has 2×2 children in the lower scale of wavelet sub-bands
SOT-B	All tree roots have their four children in the same row at the next lower scale. Nodes which are not in the LL sub-band will have the 2×2 adjacent pixels as their children, i.e. only children in scale III are "flattened"
SOT-C	Tree roots and all nodes in scale III have their four children in the same row at the next lower scale. Nodes in scale II will have the 2×2 adjacent pixels as their children, i.e. only children in scale II and scale III are "flattened"
SOT-D	All nodes have four "flattened" children at the next lower scale except at the lowest scale where the nodes will not have any descendant

parent–children relationship changes from Eqs. (6.5a) to (6.5b). The differences in these tree structures are shown in Table 6.2.

Parent at coordinate   Children at coordinate

$$(i, j) \quad \rightarrow \quad (x, y), (x, y + 1), (x + 1, y), (x + 1, y + 1)$$

$$\text{(6.5a)}$$

$$(i, j) \quad \rightarrow \quad (x, y), (x, y + 1), (x, y + 2), (x, y + 3)$$

$$\text{(6.5b)}$$

Figure 6.13 shows the original tree structure (SOT-A) and the proposed new tree structures (SOT-B, SOT-C and SOT-D) for a three-scale SOT decomposition and Table 6.3 shows the parent–children relationship at each scale for all sub-bands. This concept can be extended to cater for more scales of decomposition for all sub-bands. For example, for a five-scale decomposition, another two proposed new tree structures SOT-E and SOT-F can be obtained.

6.3.1.3   Memory Requirement with New SOT for Strip-Based Processing

Line-based wavelet transformation [32] which is adopted by the strip-based coding technique [14] performs horizontal filtering on an image in a row-by-row basis. These image rows are then stored in a temporary buffer. Vertical filtering is subsequently performed as soon as there is enough image rows stored in the buffer. This allows a low-memory implementation of DWT. For an N-scale DWT decomposition, the number of lines that is generated at the first scale when one line is generated in the Nth scale is equal to $2^{(N-1)}$. This means that the lowest memory requirement that we can achieve with an N-scale wavelet decomposition is $2^{(N-1)}$ lines. However, computation on memory requirements for wavelet-based coding algorithm needs to consider the memory needed for both the wavelet transformation and the coding algorithm.

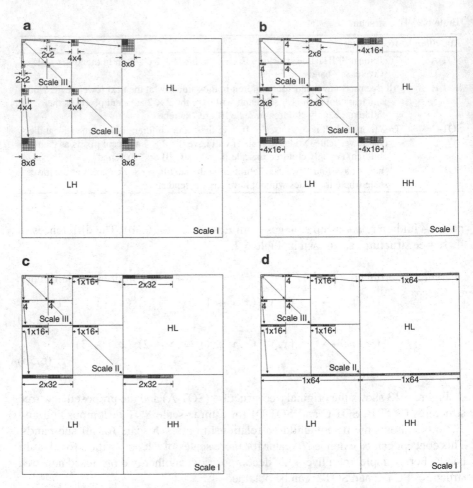

Fig. 6.13 New spatial orientation tree structures. (**a**) SOT-A, (**b**) SOT-B, (**c**) SOT-C and (**d**) SOT-D

In the SPIHT implementation described in [111], a six-scale DWT and a seven-scale SOT decomposition were performed on an image of size 512×512 pixels. Thus, the number of lines that is needed at the first scale is 128 since each node has 2×2 offspring at the lower scale and a total of seven-scale SOT decompositions were performed. The extra scale of SOT decomposition performed on the actual LL sub-band is referred to as the virtual SOT decomposition and the new LL sub-band is referred to as the virtual LL sub-band. In comparison, the strip-based SPIHT algorithm proposed in [14] was implemented with zerotree roots starting from the highest HL, LH and HH sub-bands in order to achieve minimum memory requirements at both the DWT and the encoding modules. The implementation was performed with a six-scale DWT and a six-scale SOT decomposition. Although the method proposed by [14] reduces the memory requirement by 75 % as compared

Table 6.3 Parent–children relationship of the original and the proposed new spatial orientation tree (SOT) structures for a three-scale SOT decomposition. Parent node is at coordinate (i, j) and its first descendant node is at coordinate (x, y).

Structure	Scale III	Scale II	Scale I
Parent–children relationship at HL sub-band			
SOT-A	$(x, y) = (i, j + j_R - 1)$	$(x, y) = (2i, 2j)$	$(x, y) = (2i, 2j)$
SOT-B	$(x, y) = \begin{cases}(i, 2j + j_R - 2) & \text{if } j < \frac{j_R}{2} \\ (i+1, 2j-2) & \text{otherwise}\end{cases}$	$(x, y) = (2i, 2j)$	$(x, y) = (2i, 2j)$
SOT-C	$(x, y) = \begin{cases}(i, 2j + j_R - 2) & \text{if } j < \frac{j_R}{2} \\ (i+1, 2j-2) & \text{otherwise}\end{cases}$	$(x, y) = \begin{cases}(2i, 4j + 2j_R) & \text{if } j < \frac{3j_R}{2} \\ (2i+1, 4j - 4j_R) & \text{otherwise}\end{cases}$	$(x, y) = (2i, 2j)$
SOT-D	$(x, y) = \begin{cases}(i, 2j + j_R - 2) & \text{if } j < \frac{j_R}{2} \\ (i+1, 2j-2) & \text{otherwise}\end{cases}$	$(x, y) = \begin{cases}(2i, 4j + 2j_R) & \text{if } j < \frac{3j_R}{2} \\ (2i+1, 4j - 4j_R) & \text{otherwise}\end{cases}$	$(x, y) = \begin{cases}(2i, 4j - 4j_R) & \text{if } j < 3j_R \\ (2i+1, 4j - 8j_R) & \text{otherwise}\end{cases}$
Parent–children relationship at LH sub-band			
SOT-A	$(x, y) = (i, i_R - 1, j)$	$(x, y) = (2i, 2j)$	$(x, y) = (2i, 2j)$
SOT-B	$(x, y) = \begin{cases}(i, i_R - 1, 2j) & \text{if } j < \frac{j_R}{2} \\ (i + i_R, 2j - j_R) & \text{otherwise}\end{cases}$	$(x, y) = (2i, 2j)$	$(x, y) = (2i, 2j)$
SOT-C	$(x, y) = \begin{cases}(i, i_R - 1, 2j) & \text{if } j < \frac{j_R}{2} \\ (i + i_R, 2j - j_R) & \text{otherwise}\end{cases}$	$(x, y) = \begin{cases}(2j, 4j) & \text{if } j < \frac{j_R}{2} \\ (2i+1, 4j - 2j_R) & \text{otherwise}\end{cases}$	$(x, y) = (2i, 2j)$
SOT-D	$(x, y) = \begin{cases}(i, i_R - 1, 2j) & \text{if } j < \frac{j_R}{2} \\ (i + i_R, 2j - j_R) & \text{otherwise}\end{cases}$	$(x, y) = \begin{cases}(2j, 4j) & \text{if } j < \frac{j_R}{2} \\ (2i+1, 4j - 2j_R) & \text{otherwise}\end{cases}$	$(x, y) = \begin{cases}(2j, 4j) & \text{if } j < j_R \\ (2i+1, 4j - 4j_R) & \text{otherwise}\end{cases}$

(continued)

Table 6.3 (continued)

Structure	Scale III	Scale II	Scale I
Parent–children relationship at HH sub-band			
SOT-A	$(x,y) = (i + i_R - 1, j + j_R - 1)$	$(x,y) = (2i, 2j)$	$(x,y) = (2i, 2j)$
SOT-B	$(x,y) = \begin{cases} (i + i_R - 1, 2j + j_R - 2) & \text{if } j < \frac{j_R}{2} \\ (i + 1_R, 2j - 2) & \text{otherwise} \end{cases}$	$(x,y) = (2i, 2j)$	$(x,y) = (2i, 2j)$
SOT-C	$(x,y) = \begin{cases} (i + i_R - 1, 2j + j_R - 2) & \text{if } j < \frac{j_R}{2} \\ (i + 1_R, 2j - 2) & \text{otherwise} \end{cases}$	$(x,y) = \begin{cases} (2i, 4j - 2j_R) & \text{if } j < \frac{3j_R}{2} \\ (2i + 1, 4j - 4j_R) & \text{otherwise} \end{cases}$	$(x,y) = (2i, 2j)$
SOT-D	$(x,y) = \begin{cases} (i + i_R - 1, 2j + j_R - 2) & \text{if } j < \frac{j_R}{2} \\ (i + 1_R, 2j - 2) & \text{otherwise} \end{cases}$	$(x,y) = \begin{cases} (2i, 4j - 2j_R) & \text{if } j < \frac{3j_R}{2} \\ (2i + 1, 4j - 4j_R) & \text{otherwise} \end{cases}$	$(x,y) = \begin{cases} (2i, 4j - 4j_R) & \text{if } j < 3j_R \\ (2i + 1, 4j - 8j_R) & \text{otherwise} \end{cases}$

$N = \text{number of scales of wavelet decomposition}$

$i_R = \dfrac{\text{Height of image}}{2^N}$

$j_R = \dfrac{\text{Width of image}}{2^N}$

Table 6.4 Memory requirements for the proposed work and the traditional as well as the strip-based SPIHT coding using the original tree structure

Coding scheme	DWT scale	SOT scale	Minimum Memory Lines Needed (DWT/SOT)	Type of Spatial Orientation Tree (SOT) structure
SPIHT	6	7	32/128	Original 2 × 2 SOT structure with roots at virtual LL sub-bands
Strip-based SPHIT	6	6	32/32	Original 2 × 2 SOT structure with roots starting from highest HL, LH and HH sub-bands
The proposed work	6	7	32/32	SOT-C with roots at virtual LL sub-bands
	5	6	16/16	SOT-C with roots at virtual LL sub-bands
	4	5	8/8	SOT-C with roots at virtual LL sub-bands
	4	6	8/8	SOT-D with roots at virtual LL sub-bands

to [111], its performance decreases as the number of trees is increased and many encoding bits are wasted especially at low bit rates where most of the coefficients have significant numbers of zeros. To overcome this problem, SPIHT coding can be performed using the proposed new tree structures. Table 6.4 shows the comparison in the memory requirements needed for the strip-based implementation of the proposed work using new SOT structures and those that are needed in [111] and [14].

In general, for an $N + V$ scale decomposition where N is the wavelet decomposition level and V is the further virtual SOT decomposition level, SPIHT coding using the new tree structures requires only $2^{(N-1)}$ lines of wavelet coefficients in the strip buffer. In other words, SPIHT coding using the proposed new tree structures can achieve optimal memory requirements at both the DWT module as well as the SPIHT encoding module. For example, with a four-scale DWT decomposition, the number of lines that is generated at the first scale when one line is generated at the fourth scale is equal to eight. To achieve this minimum memory requirement, either a five-scale SOT decomposition can be performed using SOT-C or a six-scale SOT decomposition can be performed using SOT-D. Therefore, by using the proposed work, the number of memory lines needed is reduced from 128 lines in [111] to only eight lines. This in effect means that the total memory required is reduced by 93.75 % when the proposed coding is used in place of the traditional SPIHT coder. It should also be noted that in the proposed work, a lower scale of DWT decomposition can be performed in conjunction with the use of the new tree structures. This not only lowers the memory requirements for the strip-based implementation but it also significantly reduces the complexity of the SPIHT coder.

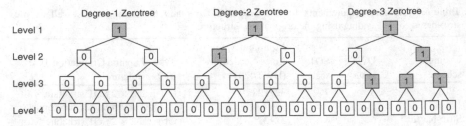

Fig. 6.14 Spatial orientation tree structure for degree-1, degree-2 and degree-3 zerotree

6.3.2   Modified SPIHT Algorithm

Zerotree coding schemes operate by exploiting the relationships among the wavelet coefficients across the different scales at the same spatial location in the wavelet sub-bands. To achieve compression, this SOT is encoded using a single symbol which indicates that all the nodes in the SOT are insignificant and there is no need to code these insignificant nodes. In general, a degree-k zerotree is a SOT where all its nodes are insignificant with respect to a threshold value except for the nodes in the top k levels. Figure 6.14 shows an example of a degree-1 to a degree-3 zerotree structure. A degree-k zerotree coder performs significance tests on a degree-1 zerotree up to a degree-k zerotree for each wavelet coefficient. From a closer examination of SPIHT coding, it can be seen that besides the individual tree nodes, SPIHT also performs significance tests on both the degree-1 and degree-2 zerotrees. Despite improving the coding performance by providing more levels of descendant information for each coefficient tested as compared to EZW which only performs significance tests on the individual tree nodes and the degree-0 zerotree, the development of SPIHT coding omits the coding of the degree-0 zerotree. Analysis from a study involving degree-k zerotree coding found that the coding of degree-0 zerotree which has been omitted during the development of SPIHT coding is important and can lead to a significant improvement in zerotree coding efficiency. Thus, in this subsection, a proposed modification to the SPIHT algorithm which reintroduces the degree-0 zerotree coding methodology will be presented. In the proposed modified SPIHT coding, significance tests that are performed on the individual tree nodes, Type A and Type B sets, are referred to as (*SIG*), (*DESC*) and (*GDESC*), respectively.

6.3.2.1   SPIHT-ZTR Algorithm

In traditional SPIHT coding, significance tests carried out on the sets in LIS are first performed on the Type A set. If the Type A set is found to be significant, i.e. $DESC(i, j) = 1$, its 2×2 offspring $(k, l) \epsilon O(i, j)$ are tested for significance and are moved to LSP or LIP, depending on whether they are significant, i.e. $SIG(k, l) = 1$ or insignificant, i.e. $SIG(k, l) = 0$, with respect to the threshold. Node (i, j) is then added back to LIS as the Type B set. Subsequently, if Type B set is found to be

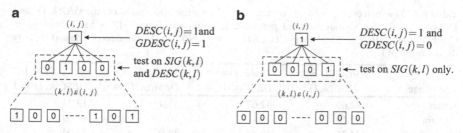

Fig. 6.15 Two combinations in modified SPIHT algorithm: (**a**) Combination 1, $DESC(i, j) = 1$ and $GDESC(i, j) = 1$; (**b**) Combination 2, $DESC(i, j) = 1$ and $GDESC(i, j) = 0$

significant, i.e. $GDESC(i, j) = 1$, the set is removed from the list and is partitioned into four new Type A subsets and these subsets are added back to LIS. In the proposed modified SPIHT algorithm, the order in which the $DESC$ and $GDESC$ bits are sent is altered. Here, the $GDESC(i, j)$ bit is sent immediately when $DESC(i, j)$ is found to be significant. As shown in Fig. 6.15, when $DESC(i, j) = 1$, four $SIG(k, l)$ bits need to be sent. However, whether the $DESC(k, l)$ bits need to be sent depends on whether $GDESC(i, j)$ is equal to "1" or "0". Thus, there are two possible combinations here: Combination 1, $DESC(i, j) = 1$ and $GDESC(i, j) = 1$; Combination 2: $DESC(i, j) = 1$ and $GDESC(i, j) = 0$.

Combination 1: $DESC(i, j) = 1$ and $GDESC(i, j) = 1$

When the significance test result of $GDESC(i, j) = 1$, it indicates that there must be at least one grand descendant node under (i, j) that is significant with respect to the current threshold T. Thus, in order to locate the significant node or nodes, four $DESC(k, l)$ bits need to be sent in addition to the four $SIG(k, l)$ bits where $(k, l) \epsilon O(i, j)$. Table 6.5 shows the results of an analysis carried out on six standard test images on the percentage of occurrence of possible outcomes of the $SIG(k, l)$ and $DESC(k, l)$ bits. As shown in Table 6.5, the percentage of occurrence of the outcome $SIG = 0$ and $DESC = 0$ is much higher than the other remaining three outcomes. Thus, in the proposed modified SPIHT coding, Huffman coding concept is applied to code all these four possible outcomes of SIG and $DESC$ bits. By allocating the least number of bits to the most likely outcome of $SIG = 0$ and $DESC = 0$, an improvement in the coding gain of SPIHT is expected. It should be highlighted that this outcome where $SIG = 0$ and $DESC = 0$ is also equivalent to the significance test of zerotree root (ZTR) in the EZW algorithm. Therefore, by encoding the root nodes and descendants of a SOT using a single symbol, the degree-0 zerotree coding methodology has been reintroduced in the proposed modified SPIHT coding, which for convenience is termed the SPIHT-ZTR coding scheme.

Combination 2: $DESC(i, j) = 1$ and $GDESC(i, j) = 0$

Table 6.5 The percentage (%) of occurrence of possible outcomes of the $SIG(k,l)$ and $DESC(k,l)$ bits for various standard greyscale test images of size 512×512 pixels under Combination 1, $DESC(i,j) = 1$ and $GDESC(i,j) = 1$. Node (i,j) is the root node and (k,l) is the offspring of (i,j)

Test image	$SIG(k,l) = 0$ $DESC(k,l) = 0$	$SIG(k,l) = 0$ $DESC(k,l) = 1$	$SIG(k,l) = 1$ $DESC(k,l) = 0$	$SIG(k,l) = 1$ $DESC(k,l) = 1$
Lenna	42.60	32.67	11.49	13.24
Barbara	42.14	35.47	10.70	11.69
Goldhill	44.76	28.13	14.07	13.04
Peppers	44.39	34.49	9.41	11.71
Airplane	44.01	25.22	16.51	14.26
Baboon	42.71	28.30	14.97	14.02
Equivalent symbol in EZW	ZTR	IZ	POS/NEG	POS/NEG
Bits assignment in the proposed work	"0"	"10"	"110"	"111"

When $DESC(i,j) = 1$ and $GDESC(i,j) = 0$, it indicates that the SOT is a degree-2 zerotree where all the grand descendant nodes under (i,j) are insignificant. It also indicates that at least one of the four offspring of node (i,j) is significant. In this situation, four $SIG(k,l)$ bits where $(k,l)\epsilon O(i,j)$ need to be sent. Let the significance of the four offspring of node (i,j) be referred to as "ABCD". Here, a total of 15 possible combinations of ABCD can be obtained as shown in Table 6.6. The percentage of occurrence of possible outcomes of ABCD is determined for various standard test images and the results are shown in Table 6.6. From Table 6.6, it can been seen that the first four ABCD outcomes of 0001, 0010, 0100 and 1000 occur more frequently as compared to the other remaining 11 possible outcomes. Like in Combination 1, Huffman coding concept is applied to encode all the outcomes of ABCD. The output bits assignment for each of the 15 possible outcomes of ABCD is shown in Table 6.6. Since fewer bits are needed to encode the most likely outcomes of ABCD, i.e. 0001, 0010, 0100 and 1000, an improved performance of the SPIHT coding is anticipated. It should be noted that in both Combinations 1 and 2, all the wavelet coefficients that are found to be insignificant are added to the LIP and those that are found to be significant are added to the LSP. The sign bit for those significant coefficients is also output to the decoder. Algorithms 6.2 and 6.3 presents the proposed modified SPIHT algorithm.

6.3.2.2 Performance of SPIHT-ZTR Coding

The performance of the proposed modified SPIHT coder is evaluated over 20 standard test images. The binary uncoded compression performance of the modified SPIHT is compared against the arithmetic encoded SPIHT coder (SPIHT-AC) and

Table 6.6 The percentage (%) of occurrence of possible outcomes of the ABCD for various standard greyscale test images of size 512 × 512 pixels under Combination 2, $DESC(i, j) = 1$ and $GDESC(i, j) = 0$. ABCD refers to the significance of the four offspring of node (i, j)

Possible outcomes of ABCD	Test images						Bit assignment in the proposed work
	Lenna	Barbara	Goldhill	Peppers	Airplan	Baboon	
0001	15.4	14.66	15.25	15.15	15.27	14.70	"00"
0010	14.87	14.21	14.41	14.76	15.84	14.67	"01"
0100	14.79	13.66	15.72	15.23	15.96	14.78	"1"+ "10"
1000	15.21	13.96	14.83	15.02	15.70	15.26	"11"
0011	4.81	5.93	5.21	5.20	5.34	5.48	"0011"
0101	5.48	5.51	5.38	4.98	4.92	4.95	"0101"
0110	4.60	4.41	4.25	4.24	3.96	4.54	"0110"
1001	4.34	4.38	4.15	4.39	3.96	4.56	"1001"
1010	5.33	5.58	5.12	5.06	5.21	4.86	"1010"
1100	4.84	5.24	5.32	5.37	5.26	5.25	"0"+ "1100"
0111	2.27	2.69	2.34	2.31	1.86	2.36	"0111"
1011	2.26	2.51	2.12	2.37	1.85	2.31	"1011"
1101	2.16	2.56	2.21	2.20	1.95	2.47	"1101"
1110	2.28	2.43	2.37	2.32	1.84	2.40	"1110"
1111	1.36	2.27	1.32	1.40	1.08	1.41	4 "1111"

the binary uncoded SPIHT coder (SPIHT-BU). In all of these simulations, a six-scale DWT and a seven-scale SOT decomposition were carried out on test images of size 512 × 512 pixels. All the three coders used the original 2 × 2 SOT-A structure. Table 6.7 presents the simulation results obtained. From the simulation results, it can be seen that the proposed modified SPIHT coder gives a better compression performance than the SPIHT-BU coder at all bit rates. On average, there is a PSNR improvement of 0.12 dB for bit rates between 0.10 and 0.75 bpp and 0.30 dB for bit rates between 1.00 and 2.00 bpp. Comparing the SPIHT-BU and SPIHT-ZTR, it can be seen that with a slight modification in the coding methodology by reintroducing the degree-0 zerotree coding in the proposed modified SPIHT coder, an appreciable increase in compression efficiency can be obtained. The simulation results also show that the binary uncoded compression performance of the modified SPIHT coder is not as good as the SPIHT-AC coder. Although the SPIHT-AC coder gives a better compression efficiency than the proposed modified SPIHT coder with an average PSNR improvement of 0.42 dB, the use of arithmetic coding not only increases the computational complexity during the implementation of the coder, it also requires additional memory for storing its multiple coding tables. Furthermore, arithmetic coding requires a longer processing time and also consumes more power during implementation. Therefore, the proposed modified SPIHT coder is more suitable for implementation in a resource constrained environment.

Algorithm 6.2 SPHIT-ZTR coding scheme

Step 1. Initialization:
Output $n = [log_2(max(i, j)|Ci, j|)]$
3: LSP is set as an empty list.
All the root nodes are added to the LIP and those nodes with descendants are added to LIS as
Type A entries.

Step 2. Sorting pass:
6: **for all** entry, (i, j) in the LIP **do**
    Output $SIG(i, j)$
    if $SIG(i, j) = 1$ **then**
9:        Move (i, j) to the LSP.
          Output the sign of $C_{i,j}$
      end if
12: **end for**
    for all entry, (i, j) in the LIS **do**
      if entry $=$ TypeA1 **then**
15:        Output $DESC(i, j)$
           if $DESC(i, j) = 1$ **then**
              go to: Loop B3 {Found in Algorithm 6.3}
18:        **end if**
      else if entry $=$ TypeA2 **then**
           Change entry (i, j) to Type A1 {The entry has been checked for significance of
           descendants in the current pass}
21:   **else if** entry $=$ TypeB1 **then**
           Output $GDESC(i, j)$
           if $GDESC(i, j) = 1$ **then**
24:           Add each $(k, l) \epsilon O(i, j)$ to the end of LIS as an entry of Type A1
              Remove (i, j) from the LIS
           end if
27:   **else if** entry $=$ TypeB2 **then**
           Change entry (i, j) to Type B1 {The entry has been checked for significance of grand
           descendants in the current pass}
      else if entry $=$ TypeB3 **then**
30:        **go to: Loop B3** {Found in Algorithm 6.3}
      end if
    end for

33: **Step 3. Refinement pass:**
    For each entry (i, j) in the LSP, except those included in the last sorting pass (i.e., with the
    same n), output the n-th most significant bit of $|Ci, j|$

    Step 4. Quantization-step update:
36: $n - 1$
    go to: Step 2. Sorting pass {Found in Algorithm 6.2 Line 5}

Algorithm 6.3 Loop B3 for SPHIT-ZTR

[LOOP B3]

Output $GDESC(i, j)$

3: **if** $GDESC(i, j) = 1$ **then**
    for all $(k, l) \epsilon O(i, j)$ **do**
        if $SIG(k, l) = 0$ and $DESC(k, l) = 0$ **then**
6:            Output 0 ($\leftarrow$ I don't understand what is to output 0. Output of what?)
        else if $SIG(k, l) = 1$ and $DESC(k, l) = 0$ **then**
            Output 10
9:        **else if** $SIG(k, l) = 0$ and $DESC(k, l) = 1$ **then**
            Output 110
        else if $SIG(k, l) = 1$ and $DESC(k, l) = 1$ **then**
12:          Output 111
        end if
        if $SIG(k, l) = 1$ **then**
15:          Add (k, l) to the LSP
          Output the sign of $C_{(k,l)}$
        else if $SIG(k, l) = 0$ **then**
18:          Add (k, l) to the end of the LIP
        end if
        if $DESC(k, l) = 1$ **then**
21:          Add (k, l) to the LIS as Type B3
        else
          Add (k, l) to the LIS as Type A2
24:          Remove (i, j) from the LIS
        end if
    end for

27: **else if** $GDESC(i, j) = 0$ **then**
    Add (i, j) to the LIS as Type B2
    Remove (i, j) from the LIS
30:    Let $ABCD$ represent the four possible outcomes of $SIG(k, l)$ where $(k, l) \epsilon O(i, j)$
    if $ABCD = 0001$ **then**
        Output 100
33:    **else if** $ABCD = 0010$ **then**
        Output 101
    else if $ABCD = 0100$ **then**
36:        Output 110
    else if $ABCD = 1000$ **then**
        Output 111
39:    **else**
        Output $0 + ABCD$
    end if

42: **end if**

Table 6.7 Performance of the proposed SPIHT with degree-0 zerotree (modified SPIHT) coding compared to arithmetic encoded SPIHT (SPIHT-AC) coding and binary uncoded SPIHT (SPIHT-BU) coding in terms of average peak signal-to-noise ratio, PSNR in decibels (dB) versus bit rate or bits per pixel (bpp) for 20 test images of size 512×512 pixels

Bitrate	Average PSNR (dB)		
(bpp)	SPIHT-AC	SPIHT-ZTR	SPIHT-BU
0.10	25.30	24.99	24.94
0.15	26.64	26.32	26.24
0.20	27.67	27.29	27.24
0.25	28.55	28.18	28.07
0.30	29.32	28.93	28.81
0.35	30.00	29.63	29.49
0.40	30.61	30.22	30.10
0.45	31.20	30.78	30.64
0.50	31.75	31.32	31.16
0.75	34.02	33.59	33.40
1.00	35.91	35.44	35.22
1.25	37.58	37.10	36.83
1.50	39.14	38.62	38.30
2.00	42.07	41.48	41.10

6.3.3   Listless SPIHT-ZTR Coding

Although the proposed SPIHT-ZTR coding scheme which uses a degree-0 to a degree-2 zerotree coding methodology can be implemented using the set partitioning approach as described in Algorithms 6.2 and 6.3, its implementation in a hardware constrained environment is difficult. One of the major difficulties encountered is the use of three lists to store the coordinates of the individual coefficients and subset trees during the set partitioning operation. These lists are variable in length, are data dependent and require memory management since the nodes are moved, added or removed during the sorting pass. This increases not only the complexity of the coder but also the cost of the hardware implementation as a large amount of storage is needed to maintain the lists. The listless SPIHT-ZTR coding is essentially a modified SPIHT-ZTR algorithm where the use of lists has been eliminated. A listless coder greatly reduces the memory storage needed during the image coding process. The proposed listless algorithm not only has all the advantages that a listless coder has, it is also specially developed for the low-memory strip-based implementation of the SPIHT-ZTR coding.

The encoding path of each wavelet coefficient using the SPIHT-ZTR algorithm is examined and analysed. Following this, a flow chart of the proposed listless SPIHT-ZTR coding is developed and presented in Fig. 6.16 to Fig. 6.18. In the proposed listless SPIHT-ZTR algorithm, significance tests performed on an individual tree node, descendant of a tree node and grand descendant of a tree node are referred

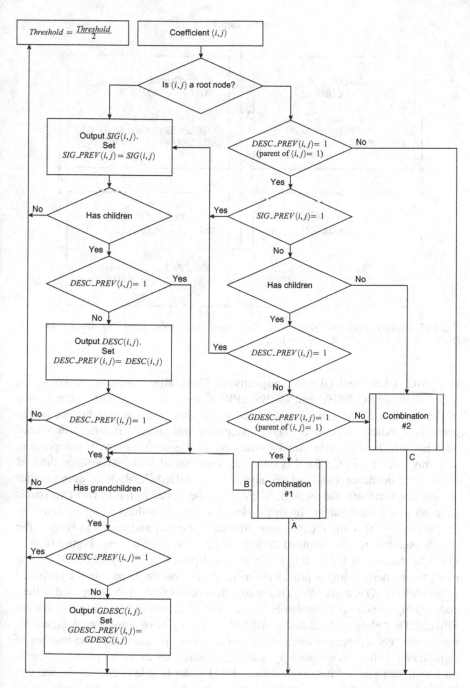

Fig. 6.16 Listless modified SPIHT for strip-based implementation. (**a**) Sorting pass and refinement pass

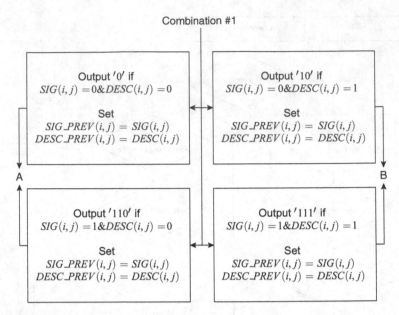

Fig. 6.17 Listless modified SPIHT for strip-based implementation. (**b**) Combination 1: $DESC(i, j) = 1$ and $GDESC(i, j) = 1$

to as SIG, DESC and GDESC, respectively. Three significance maps known as SIG_PREV, DESC_PREV and GDESC_PREV are used to store the significance of the coefficient, the significance of the descendant and the significance of the grand descendant, respectively. The SIG_PREV information is stored in a one-bit array which has a size equal to the size of the strip buffer. In comparison, the array size of DESC_PREV is only a quarter that of SIG_PREV since the leaf nodes have no descendant and the array size of GDESC_PREV is even smaller at only one sixteenth that of SIG_PREV since the nodes in the lowest two scales have no grand descendant. In strip-based coding, after the coding is done for an image strip, this significance map storage is cleared and ready to receive the significance test results obtained during the next round of coding. Figure 6.16 to Fig. 6.18 shows that the sorting and refinement passes in the SPIHT-ZTR coding using the set partitioning approach are merged into one single pass in the proposed listless SPIHT-ZTR algorithm. This makes the control flow of the proposed listless coding simple and easy to implement in hardware. Even though the proposed listless SPIHT-ZTR coding and the traditional SPIHT coding have similar peak signal-to-noise ratio (PSNR) performance at the end of every bit plane since the number of significant coefficients encoded by both algorithms after every bit plane is exactly the same, the proposed listless modified SPIHT coder is able to encode the wavelet coefficients more efficiently as it requires fewer bits to encode the same set of wavelet coefficients compared to the traditional SPIHT coder.

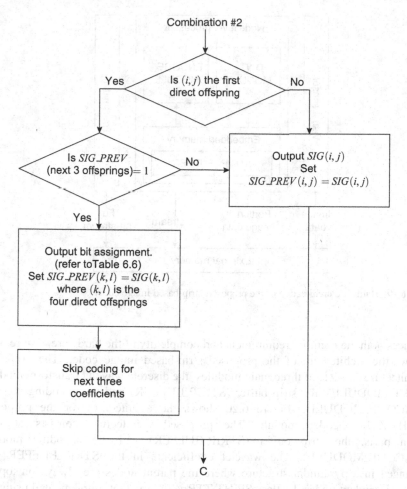

Fig. 6.18 Listless modified SPIHT for strip-based implementation. (**c**) Combination 2: $DESC(i, j) = 1$ and $GDESC(i, j) = 0$

6.4 Hardware Implementation on the WMSN Platform

In this section, hardware implementation of a listless SPIHT-ZTR coding on the WMSN platform is discussed. For low-memory implementation, the image coder uses the SPIHT-ZTR algorithm based on a degree-0 to degree-2 zerotree coding methodology to achieve a better compression performance without the need for adaptive arithmetic coding which would otherwise require extra memory to store the multiple coding tables. In addition, to facilitate the hardware implementation, the SPIHT-ZTR coding uses the SOT-B tree structure and eliminates the use of lists in its set partitioning approach and is implemented using the upward–downward scanning methodology. This listless approach and upward–downward scanning methodology

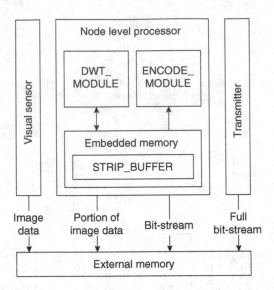

Fig. 6.19 Hardware architecture of the proposed strip-based image coder

reduces both the memory requirement and complexity of the hardware. Figure 6.19 shows the architecture of the proposed strip-based image coder. The proposed architecture consists of three main modules: the discrete wavelet transform module (DWT_MODULE), the strip buffer (STRIP_BUFFER) and the encoding module (ENCODE_MODULE). Figure 6.20 shows the architecture for the proposed SPIHT-ZTR encoder module. The proposed architecture consists of two main parts: the strip buffer (STRIP_BUFFER) and the encoding module (ENCODE_MODULE). The wavelet coefficients in the STRIP_BUFFER are arranged in a pyramidal structure where the parent nodes are always on top of their descendant nodes. Listless SPIHT-ZTR coding is then implemented using an upward–downward scanning methodology.

To facilitate the SPIHT coding, the wavelet coefficients obtained from the discrete transform module are stored in the STRIP_BUFFER at predetermined locations. Figure 6.21 shows a memory allocation example of the wavelet coefficients in the STRIP_BUFFER. Here, assuming a three-scale DWT decomposition is performed on an image strip of size 8×128 pixels. As shown in Fig. 6.21, a total of 16 root nodes are obtained during the initialization stage of the SPIHT coding, i.e. four roots that have no descendant and the other 12 roots that are the roots for the high–low (HL), low–high (LH) and high–high (HH) sub-bands. It should be noted that the wavelet coefficients in the strip buffer are arranged in such a manner that each parent node will have its four direct offspring in a consecutive order. Since the wavelet coefficients in the STRIP_BUFFER are arranged in a pyramidal structure where the parent nodes are always on top of their descendant nodes, a listless SPIHT-ZTR coding is implemented using an upward–downward scanning methodology.

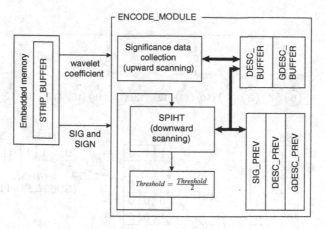

Fig. 6.20 Architecture for SPIHT-ZTR encoder module

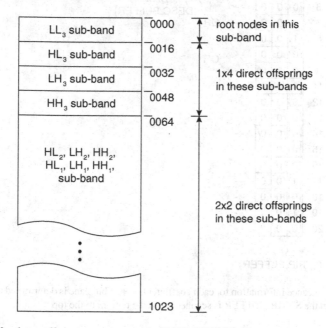

Fig. 6.21 Wavelet coefficients arrangement in STRIP_BUFFER

6.4.1   Stages 0 and 1: Data Copy and DWT

The hardware implementations for Stages 0 and 1 (Data Copy and DWT) have been described in Sects. 4.4.3 and 4.4.4, respectively.

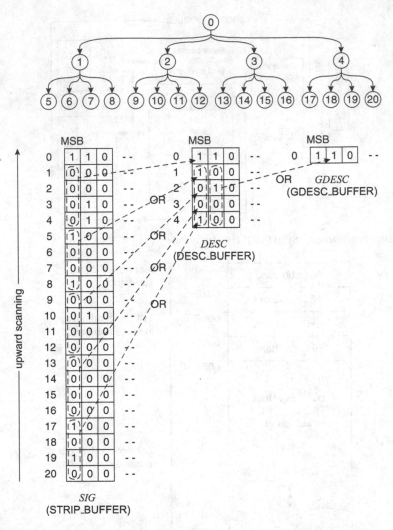

Fig. 6.22 Significance information for each coefficient at each bit plane is determined and is stored in buffers when the STRIP_BUFFER is scanned from the bottom to the top

6.4.2 Stage 2: Upward Scanning Significance Data Collection

This scanning method starts from the leaf nodes up to the roots, i.e. from the bottom to the top of the STRIP_BUFFER. While the zerotree is being scanned, the *DESC* and *GDESC* significance information for each coefficient at each bit plane is determined and stored in temporary buffers DESC_BUFFER and GDESC_BUFFER. This significance data collection process is carried out in parallel for all bit planes as shown in Fig. 6.22. The *SIG* information is obtained directly from the

---

Algorithm 6.4 Upward scanning: significance data collection

Assume:
STRIP_RAM S1 partition holds the image strip
3: STRIP_RAM S2 partition holds the DESC_BUFFER
STRIP_RAM S3 partition holds the GDESC_BUFFER
$S1(X)$ means the 16 bits of data at location X of S1

6: **Build the DESC significance table:**
$R5 = sizeofS1partition$
$R6 = 1$
9: $R7 = R5 - 4$

while $R7 < R6$ **do**
for all bits in the location S1 **do**
12:        $R8 =$
$S1(R7)$ **or**
$S1(R7 + 1)$ **or**
15:        $S1(R7 + 2)$ **or**
$S1(R7 + 3)$
$S2(R7) = R8$
18:        $R7 = R7 - 4$
end for
end while

21: **Build the GDESC significance table:**
$R9 = sizeofS2/4$
$R10 = 1$
24: $R11 = R9 - 4$

while $R11 >= R10$ **do**
for all bits in the location S2 **do**
27:        $R12 =$
$S2(R11)$ **or**
$S2(R11 + 1)$ **or**
30:        $S2(R11 + 2)$ **or**
$S2(R11 + 3)$
$S3(R7) = R12$
33:        $R11 = R11 - 4$
end for
end while

STRIP_BUFFER, whereas the *DESC* and *GDESC* information for a coefficient is obtained by OR-ing the *SIG* and *DESC* results of its offspring, respectively. With all the significance information pre-computed and stored, the encoding process is speeded up since the significance information can be readily obtained from the buffers during SPIHT coding. Algorithm 6.4 shows the processing step for the significance data collection.

6.4.3   Stage 3: Downward Scanning Listless SPIHT-ZTR Coding

Listless modified SPIHT-ZTR coding is performed on the wavelet coefficients stored in the STRIP_BUFFER. An embedded coding can also be obtained here by performing a multiple-pass downward scanning methodology on the wavelet coefficients stored in the STRIP_BUFFER. During the downward scanning process, the wavelet coefficients are encoded in the order of importance, i.e. those coefficients with a higher magnitude are encoded first. Algorithms 6.5–6.8 show the processing step for the multiple-pass listless SPIHT-ZTR coding. It should be noted that these algorithms are derived from flow chart presented in Fig. 6.16.

In the case of a non-embedded coding, it can be performed using the one-pass downward scanning methodology. Here, instead of scanning the spatial orientation trees for different magnitude intervals, each coefficient in the tree is scanned starting from its most significant bit (MSB) to the least significant bit (LSB). Since all the significance information needed for all bit planes is stored during the upward scanning process, a full bit-plane encoding can be carried out on one coefficient followed by the next coefficient. As each coefficient in the STRIP_BUFFER is only accessed once during encoding, the one-pass downward scanning process reduces the frequency of memory access during the SPIHT-ZTR encoding stage.

6.5   Hardware Setup and Results

This section describes a practical implementation of the strip-based SPIHT-ZTR compression for a human face detection and tracking system in a WMSN environment. In the proposed implementation, image data captured by a visual sensor node is first compressed using a low-memory strip-based listless SPIHT-ZTR compression scheme using the SOT-C structure. After compression, the encoded bitstream is transmitted to the network motes which are used to relay the encoded data throughout the WMSN until it reaches the base station which interfaces the WMSN to a host workstation. At the host workstation, the received bitstream is decoded and the reconstructed image sequence is then fed into a face detection and tracking system, which is implemented using the Viola–Jones (VJ) algorithm [130]. Figure 6.23 shows the experimental setup of the proposed implementation of visual compression in a face detection and tracking system in a WMSN. The WMSN is implemented with the Celoxica RC203E FPGA platform.

As shown in Fig. 6.24, the back-end processing is carried out in two stages: the decompression stage and the face detection and tracking stage. In the proposed implementation, a decompression module is developed in MATLAB. This software decoder uses a similar architecture flow as the hardware encoder presented in

---

Algorithm 6.5 Downward scanning: listless SPIHT-ZTR coding

Initial:
STRIP_RAM S1 partition holds the image strip.
3: STRIP_RAM S2 partition holds the DESC_BUFFER.
STRIP_RAM S3 partition holds the GDESC_BUFFER.
STRIP_RAM S4 partition holds the SIG_PREV (S4'3), DESC_PREV (S4'2), and
                    GDESC_PREV (S4'1) where each of them is stored at bit 3, bit 2 and bit 1
                    in each location of S4, respectively.
6: $S1(X)$ means the 16 bits of data at location X of S1.
$S1(X, Y)$ means the $(16 - Y)$th bit of data at location X of S1.

Multiple-pass downward scanning:
9: $R13 = numberofplane$
$R14 = bitplane = 0$
$R15 = sizeofS1$
12: $R16 = coefficient = 0$

for $R14 <= $R13 do
    for $R14 <= $R13 do
15:     if $R16isarootnode then
            go to: 6.13{This Algo does NOT exist!}
        else if S4'2(parentof$R16) == 1 then
18:         if S4'3($R16) == 0 then
                if $R16hasadescendent then
                    if S4'2($R16) == 1 then
21:                     go to: Algorithm 6.6 {Downward Scanning: Encode Significance Informa-
                        tion}
                    else if S4'1(parentof$R16) == 1 then
                        go to: 6.4{Downward Scanning: Combination 1}
24:                 else
                        go to: 6.5{Downward Scanning – Combination 2}
                    end if
27:             end if
            else
                go to: 6.3{Downward Scanning: Encode Significance Information}
30:         end if
        end if
        $S16 = $S16 + 1
33: end for
    $R14 = $R14 + 1
end for

Sect. 6.4. Here, a bitstream received from the hardware compression module is decoded by the software decoder and image sequences captured by the camera are reconstructed. The face detection and tracking system using the Viola–Jones (VJ) algorithm provided by [130] is adopted in this implementation. The program runs in MATLAB using the OpenCV library. The experiments for the implementation of the proposed strip-based visual compression in the human face detection and tracking system in a WMSN were conducted under line-of-sight conditions performed outdoors as well as in an indoor non-line-of-sight environment. Figures 6.25 and

---

Algorithm 6.6 Downward scanning: encode significance information

---

    Assume:
    STRIP_RAM S1 partition holds the image strip
3: STRIP_RAM S2 partition holds the DESC_BUFFER
    STRIP_RAM S3 partition holds the GDESC_BUFFER
    STRIP_RAM S4 partition holds the SIG_PREV (S4'3), DESC_PREV (S4'2), and
                  GDESC_PREV (S4'1) where each of them is stored at bit 3, bit 2 and bit 1
                  in each location of S4, respectively
6: S1(X) means the 16 bits of data at location X of S1.
    S1(X, Y) means the $(16 - Y)$th bit of data at location X of S1.
    $R14 = $ currentbit $-$ planecodingfromAlgo. 6.5
9: $R16 = $ currentcoefficientlocationfromAlgo. 6.5
    binary encode S1($R16, $R14)
    set S4'3 $=$ S1($R16, $R14)

12: **if** $R16 has descendents **then**
      if S4'2($S16) $==$ 0 **then**
          binary encode S2($R16, $R14)
15:        **set** S4'2 $=$ S2($R16, $R14)
      end if
    end if

18: **if** $R16 has grand descendents **then**
      if S4'3($S16) $==$ 0 **then**
          binary encode S3($R16, $R14)
21:        **set** S4'3 $=$ S3($R16, $R14)
      end if
    end if

6.26 show the test results obtained from the experiments carried out in both outdoor and indoor conditions, respectively. It should be noted that for the experiments carried out in an indoor environment, the camera, encoding module and the XBee ZNet 2.5 transmitter were placed inside a room, whereas the XBee ZNet 2.5 receiver and the host workstation were placed inside another room. In these experiments, image sequences are first captured by the camera and stored into the ZBT RAM provided with the RC203E board. These image sequences are subsequently strip-based encoded. The bitstream generated from the compression module is then stored back into the ZBT RAM. Upon request from the back-end processing side, the compressed bitstream is transmitted through the XBee ZNet 2.5 RF modules to the host workstation. At the host workstation, the bitstream is decoded and the reconstructed image sequences are then fed into the face detection and tracking system. In these experiments, all image sequences were reconstructed at decoding bit-plane number six. As shown in Figs. 6.25 and 6.26, faces of several people were detected and tracked accurately in the proposed strip-based visual compression in the human face detection and tracking system in a WMSN environment.

Algorithm 6.7 Downward scanning: combination 1

   $S1(X)$ means the 16 bits of data at location X of S1.
   $S1(X, Y)$ means the $(16 - Y)$th bit of data at location X of S1.
3:   $\$R14$ = currentbit − planecodingfromAlgo. 6.5
     $\$R16$ = currentcoefficientlocationfromAlgo. 6.5
     $\$R20$ = S1($\$R16, \$R14$)
6:   $\$R21$ = S2($\$R16, \$R14$)

     if $\$R20 == 0\&\$R21 == 0$ **then**
        binary encode "0"
9:      set S4$'$3 = $\$R20$
        set S4$'$2 = $\$R21$
     else if $\$R20 == 0\&\$R21 == 1$ **then**
12:     **binary encode** "10"
        set S4$'$3 = $\$R20$
        set S4$'$2 = $\$R21$
15:     **if** $\$R16$ has grand descendents **then**
           if S4$'$1($\$S16$) $== 0$ **then**
              set S4$'$1 = S3($\$R16, \$R14$)
18:        **end if**
        end if
     else if $\$R20 == 1\&\$R21 == 0$ **then**
21:     **binary encode** "110"
        set S4$'$3 = $\$R20$
        set S4$'$2 = $\$R21$
24: **else if** $\$R20 == 1\&\$R21 == 1$ **then**
        binary encode "111"
        set S4$'$3 = $\$R20$
27:     set S4$'$2 = $\$R21$
        if $\$R16$ has grand descendents **then**
           if S4$'$1($\$S16$) $== 0$ **then**
30:           **binary encode** S3($\$R16, \$R14$)
              set S4$'$1 = S3($\$R16, \$R14$)
           end if
33:     **end if**
     end if

Algorithm 6.8 Downward scanning: combination 2

   S1(X) means the 16 bits of data at location X of S1.
   S1(X, Y) means the $(16 - Y)$th bit of data at location X of S1.
3: $R14 = currentbit − planecodingfromAlgo. 6.5
   $R16 = currentcoefficientlocationfromAlgo. 6.5

   if $R16isthefirstdirectoffspring **then**
6:    **if** §4'3(next3offspring) $== 0$ **then**
         Output bit assignment based on Table 6.6
         S4'3 (direct4offspring) $=$ S1 (direct4offspring)
9:       $R16 = $R16 + 3
      else
         binary encode S1($R16, $R14)
12:      **set** S4'3 $=$ S1($R16, $R14)
      end if
   end if

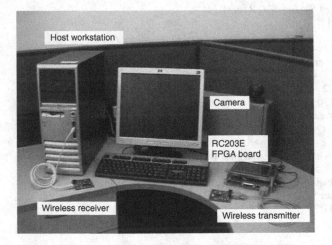

Fig. 6.23 Experimental setup of the proposed implementation of visual compression in a face detection and tracking system in a wireless multimedia sensor network

Fig. 6.24 Back-end processing. (**a**) Decompression stage. (**b**) Face detection and tracking stage

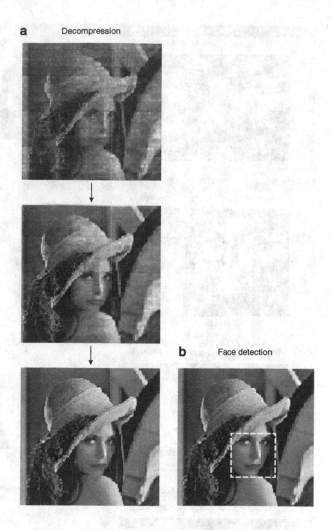

Fig. 6.25 Experimental results of strip-based visual compression in the human face detection and tracking system under line-of-sight *outdoor* condition

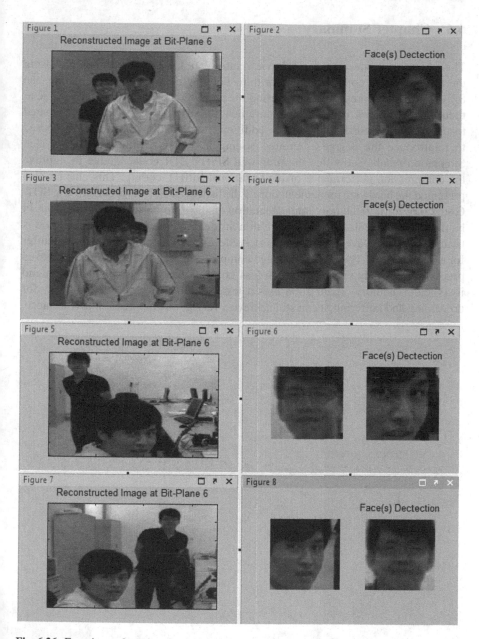

Fig. 6.26 Experimental results of strip-based visual compression in the human face detection and tracking system under line-of-sight *indoor* condition

6.6   Chapter Summary

This chapter introduced the single-view information reduction techniques using the SPIHT image compression algorithm. For low-memory implementation, the SPIHT image compression scheme using the proposed new SOT structure is able to reduce up to 93.75 % of memory requirement for strip-based processing compared to the traditional SPIHT coding. In addition, modifications have been proposed to reintroduce the degree-0 zerotree coding into the traditional SPIHT coder. Simulation results show that the proposed SPIHT-ZTR coder is able to improve the compression performance of the traditional SPIHT coding. This chapter also presented the hardware implementation of the listless SPIHT-ZTR coder using SOT-C structure for low-memory implementation on the strip-based MIPS processor. In hardware implementation, a new one-dimensional, memory-addressing method is used to store the wavelet coefficients at predetermined locations in the strip buffer for ease of coding. The proposed image compression architecture using strip-based processing not only requires a much less complex hardware implementation and its efficient memory organisation uses a lesser amount of embedded memory for processing and buffering, it can still achieve a very good compression performance.

Chapter 7
Multiple-View Information Reduction Techniques for WMSN Using Image Stitching

Abstract The wireless multimedia sensor network (WMSN) is an example of a platform for a multiple camera application. Since the devices in the WMSN are powered by batteries, energy consumption remains a critical issue. The previous chapters have discussed methods for single-view information reduction by removing the redundancy present in an image using event compression or by removing unimportant image frames using event detection. For multiple camera applications, additional redundancy can be found in cases where there is an overlap in the field of view (FOV) of different nodes. This is considered as the inter-camera redundancy that can be exploited to further reduce the data requirements for transmission. In this chapter, several approaches that have been proposed for multiple-view information reduction will be reviewed. The chapter begins by discussing approaches to remove or reduce the inter-camera redundancy. This is followed by a discussion on feature extraction which is a core technique used in searching for the similarities between two images. Then, an overview of the image stitching process will be provided. The performance of using different feature extraction methods to locate the same features in small size images and perform the stitching will be presented as well. The chapter concludes with an implementation of the proposed framework on the WMSN hardware platform.

7.1 Introduction

Although there is always a trade-off in sensor nodes between the energy required for processing and the energy required for transmission, it has been shown in [137] that the energy required to transmit the data is higher than the energy required to process the same amount of data. Due to this reason, it is better to focus on reducing the amount of data that need to be transmitted. For multiple camera applications, this can be realised by exploiting the data redundancy from the different cameras.

L. Ang et al., *Wireless Multimedia Sensor Networks on Reconfigurable Hardware*,
DOI 10.1007/978-3-642-38203-1_7, © Springer-Verlag Berlin Heidelberg 2013

The algorithms and architectures that are designed to achieve this should fulfil the three requirements as follows:

1. Low-complexity algorithms and hardware
2. High coding performance
3. No need for precise camera calibration

The first requirement is essential because the sensor nodes have limited battery power and hardware resources. In order to reduce the energy required for data transmission, image compression techniques can be applied. Even though the implementation of compression algorithms would increase the computational complexity and lead to higher energy consumption, it can significantly reduce the energy used for transmission. According to [85], the energy required to transmit 1 kbit of data for 100 m is similar to the energy taken to process 3 million of instructions in a general purpose processor. This also matches with the findings in [25] and [49] which also show that the energy used for transmission is much higher than the energy used for processing. This leads to the second requirement and the need for algorithms that can achieve high coding performance within the hardware constrained environment. The third requirement is specific for multi-camera applications such as the WMSN. It is highly desirable that the visual sensor nodes do not require precise placement during installation or operation. Ideally, it would be preferable to be able to just drop the nodes within the WMSN and expect them to operate without the need for further calibration on the cameras. In this chapter, a framework that attempts to fulfil the requirements by combining image stitching and compression is proposed. Image stitching is an image processing technique that is used to combine multiple images with overlapping FOVs into the same coordinate system as shown in Fig. 7.1. This technique can also be used to enlarge the FOV of an image.

In the proposed framework, each visual node is considered to be only equipped with a single image sensor, but many visual nodes can be deployed to provide different viewpoints of the scene. In order to reduce the burden on the sensor node, the process of locating the overlapping regions with image stitching is shifted to the aggregation node. The aggregation node can be a node with higher computational capability or the host workstation. By doing this shifting, the computations that are not favourable for hardware implementations, such as floating-point multiplication and division operations, will be performed at the aggregation node. A small amount of parameters is generated, and they can be used by visual nodes to remove the redundancy created by the overlapping regions.

7.2 Joint-Sources Compression

Joint-sources compression is another name to address the approach that is capable to remove the inter-camera redundancy. The development of the joint-sources compression can be divided into two categories depending on whether the redundancy is removed at the encoder or at the decoder. In the first category, the frameworks are

Fig. 7.1 The effect of applying image stitching to images captured by two image sensors

mainly developed based on the Slepian–Wolf (SW) theorem [115] or the Wyner–Ziv (WZ) theorem [138]. These methods are based on prediction. Consider the case that there are two correlated sources denoted as X and Y, respectively. The decoder will try to predict the data transmitted by Y based on the data that has been received from X. Theoretically, Y is only required to send a small amount of data to correct the errors that might have occurred during the reconstruction. This helps to shift the computational burden to a centralised decoder and allows a simple encoder to be constructed. In order to predict the data from Y precisely, it is required to model the correlation between X and Y. However, it is difficult to model the correlation accurately, and this is one of the main reasons that lead to the performance gap for practical implementations [53].

The frameworks from the second category avoid relying on the prediction. The technique adopted in these frameworks can be denoted as collaborative correspondence analysis. The basic idea of this is to first perform an analysis on a set of images captured by different image sensors and search for the redundancy. In this case, it is required for those correlated sources to exchange information among each other. This will increase the computational burden and is less favourable for the visual node. However, these frameworks are more reliable since no prediction is required. The following subsections will discuss several frameworks that fall in the first and the second category.

7.2.1 Slepian–Wolf and Wyner–Ziv Theorems

The main concept of the SW theorem and the WZ theorem is about the same. Just that the former case is for lossless compression, while the latter case is for lossy compression. One of the main reasons for using the SW or WZ theorems is to reduce the computational burden on the encoder by shifting them to the decoder. In conventional compression such as MPEG2, the complexity of the encoder is higher than the decoder. The encoder is required to perform an extensive amount of computation that involves the process of transformation, coding, motion estimation and motion compensation. The decoder is only required to decode the incoming bitstream and the motion vectors. This condition is not ideal for the implementation in the WMSN due to the energy and hardware constraints on the visual node. The SW and WZ theorems can fulfil the requirement of the WMSN by shifting the computational burden from the encoder to the decoder.

Assuming that X and Y are two separate sources with certain correlation, the compression of X with the presence of Y at the encoder and decoder is illustrated in Fig. 7.2a. According to the SW or WZ theorems, it is possible to achieve the same compression rate with only Y present at the decoder as depicted in Fig. 7.2b. The same compression rate can be achieved without having to exchange information between X and Y at the encoder. In this case, side information that is generated from Y is first produced at the decoder. It is the prediction of X from Y and it will be used to aid the process of reconstructing X. This shifts the computational burden from the encoder to a centralised decoder. However, the performance of these methods is strongly affected by the accuracy of the side information. The side information needs to be predicted as similar to X as possible to achieve a good performance.

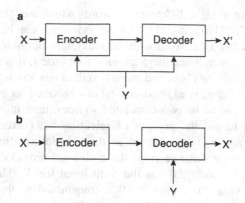

Fig. 7.2 (a) The encoding of X with Y present at both encoder and decoder. (b) The encoding of X with Y only present at the decoder

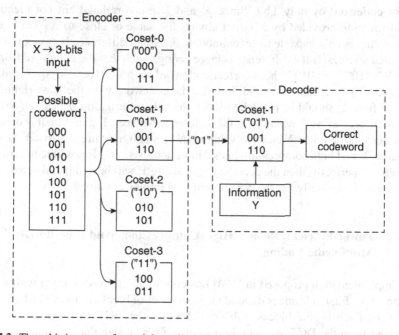

Fig. 7.3 Three-bit input can form eight codewords, and these eight codewords can be separated into four different cosets

7.2.1.1 Distributed Source Coding Using Syndromes

This framework, which is proposed in [103], evaluated and modelled the use of coset coding to meet the SW and WZ theorems. The basic idea of coset coding is shown with an example given in Fig. 7.3. In this case, X is assumed to be a 3-bit input that will be encoded with only Y present at the decoder. For a 3-bit input, it can form a maximum of eight different codewords which are "000", "001", "010", "011", "100", "101", "110" and "111". These codewords can be divided into four different cosets. The two codewords with the maximum Hamming distance will be placed into the same coset. Since there are only four cosets, it is sufficient to index the four cosets with two data bits, and these two data bits are denoted as the index bits. When the coset coding is adopted, one bit can be saved for every 3-bit input.

If X and Y are said to be two correlated sources, then information provided by Y can be used to aid the process of selecting the correct codeword from the coset. In this case, Y is considered as the side information. The codeword with the least Hamming distance from the codeword provided by Y should be selected. For example, take "001" as the 3-bit input for X. This codeword falls in the coset index with "01", thus, "01" is transmitted to the decoder. In this case, if the codeword from Y is "001" as well, then the outcome of the selection is identical. However, even if the codeword from Y is "011", "101" or "000", the correct codeword can still be picked because these codewords differ from the correct codeword by only 1 bit. Since X and Y are correlated but not identical, the information provided by Y is not always the same or close to X. The use of coset coding is only capable to accommodate the small difference between the two correlated sources. If the difference is large enough such that the codeword from Y is "110", "100" or "010", the incorrect codeword such as "100" or "010" will be selected. This is to fulfil the condition that the codeword with the least Hamming distance from Y should be picked. Overall, the side information, Y, can be written as $Y = X + N$ as well, where N is the difference between Y and X that is denoted as the correlation noise. When the value of N is small, this implies that Y is very similar to X and high compression rates can be achieved. If it is possible to estimate the value N perfectly, then the decoding of X using Y will be completely error free. However, this is usually not the case in practical implementations.

7.2.1.2 Power-Efficient, Robust, High-Compression, Syndrome-Based Multimedia Coding

This implementation proposed in [104] has combined the coset coding with video compression. First, a frame is divided into a number of blocks in the size of 16×16 or 8×8. Each of the blocks will pass through the DCT, and the coefficients generated from the DCT are quantised by using a step size that is proportional to the correlation noise, N. As shown in an example in Fig. 7.4a, a decoding error will occur if the step size is smaller than N. In this case, the black node is the correct codeword. However, the grey node will be chosen since Y is closer to the grey node. Hence, it is necessary to use a coarser step size as shown in Fig. 7.4b.

After the step size has been decided, the next task is to partition the codeword into different cosets. The coefficients generated from the DCT will then be encoded based on the cosets to produce the output bits. Since the step size is inversely proportional to the compression rate, another quantisation will be performed on the output bits to recover the loss in performance incurred by N.

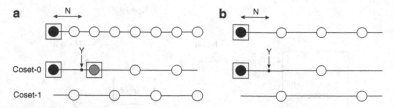

Fig. 7.4 The difference between using a (**a**) finer step size and (**b**) coarser step size

7.2.2 Collaborative Correspondence Analysis

The basic idea of collaborative correspondence analysis is to first perform an analysis on a set of images captured by different visual nodes and search for the overlapping regions. The overlapping regions can be determined by using image registration which involves the processes of extracting the feature points and then matching the feature points to determine the overlapping regions in those images. After the overlapping regions are known, a series of post-processing steps can be carried out to reduce the data that need to be transmitted. However, this will increase the computational burden on the visual nodes due to the additional processing that is required to be carried out.

7.2.2.1 Image Compression Using Correspondence Analysis and Super Resolution

In [131], a low bandwidth channel is required for the visual nodes to exchange information regarding the overlapping regions. In this case, edge detection is implemented in each of the visual nodes, and one of them is designated as the reference node. The information generated from the edge detection in the reference node will be shared among the visual nodes. It should be noted that the edge detection is applied to the subsampled version of the captured image. This is to reduce the computational burden on the visual nodes. At the same time, other visual nodes will also apply the edge detection on each of the images captured by itself. The information from the reference node is then compared to the information generated in each of the visual nodes to identify the overlapping regions. Then, the low-resolution representation of the overlapping regions will be transmitted to the centralised decoder. The super-resolution technique is used to reconstruct the full resolution of the overlapping regions by using multiple numbers of the low-resolution representation that are received from the other visual nodes. The performance of this method is proportional to the number of visual nodes that share the same FOV because the size of the low-resolution representation can be made smaller.

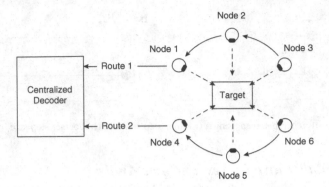

Fig. 7.5 An example of the setup of the visual nodes in the WMSN that need to go through multi-hops before reaching the centralised decoder

7.2.2.2   Collaborative Image Coding

Another similar method is proposed in [137]. In this case, a technique similar to [131] is used to find out the overlapping regions. Different from [131], this method assumes that the data will need to go through multiple hops (multi-hops) before reaching the centralised decoder as shown in Fig. 7.5. In this case, each of the visual nodes can only obtain access to the data that is transmitted from the previous node. It should be noted that the data consists of the image and a set of feature points that defines the overlapping regions. Each of the visual nodes first transmits an image to the centralised decoder once and these images are used as the background for the incoming images. Next, all the visual nodes will enter the sleeping state until an event is detected by the visual nodes.

In addition to [131] that focuses on the spatial redundancy, this method also exploits and removes the temporal redundancy. It is assumed that the FOV of the visual nodes would remain static all the time. Hence, the background of the images captured by these visual nodes should remain the same as well. Therefore, this method chooses to send only the background once at the beginning of the operation. Then for the subsequent images, only the area that surrounds the object of interest is required to be transmitted.

7.2.3   Comparison of Joint-Sources Compression

The frameworks developed based on the SW theorem or the WZ theorem are very suitable to be implemented in the WMSN. Due to the energy and hardware constraints on the visual nodes, the use of the SW theorem or WZ theorem can provide a low-complexity encoder that is more preferable, and the computational burden is shifted to the decoder. This reduces the computational burden on the visual nodes. However, the reliance on prediction has created a performance gap for practical implementations. The frameworks developed based on the collaborative correspondence analysis have tried to avoid from using prediction. Although the computational burden incurred on the visual nodes is higher, the performance gap between the theoretical and practical implementations is smaller. In the next section, feature extraction which serves as the core technique in locating the overlapping regions will be discussed in more detail.

7.3 Feature Extraction

The first task to determine the overlapping regions is to search for the similarities such as the distinct object, background or uniqueness feature in a set of images. This can be done by using either pixel-based approaches or feature-based approaches. The basic idea of pixel-based approaches is to first select a block of pixels in image A and try to search for the same block of pixels in image B. According to [119], the degree of matching can be measured by using the sum of absolute difference (SAD), mean of absolute difference (MAD), sum of square difference (SSD) or normalised cross-correlation (NCC). The computational and memory complexity of the pixel-based approaches can vary with respect to how it is implemented. For example, using the coarse-to-fine searching method can help to reduce the complexity since the searching process at the coarse level is much faster than in the fine level and the searching can be halted once a reliable result is obtained. In addition, if the searching can pass over to different levels, it is also possible to improve the performance in locating the same block even though there are changes in scale. However, this will increase the memory requirement because it is required to store a set of images at different resolutions. The pixel-based approaches are also less robust when there are rotations and changes in viewpoint [119]. Instead of looking for the same pixels, feature-based approaches will try to search for unique feature points such as corners and blobs (local maxima or minima). Many feature-based approaches are capable to overcome the problem even when there are changes in scale, rotation, illumination or viewpoint. Five approaches that are commonly used will be described and compared in the following subsections. This includes the Harris corner detector [56], Harris–Laplace detector [92], Laplacian-of-Gaussian (LoG) detector [79], features from accelerated segment test (FAST) detector [110] and scale invariant feature transform (SIFT) detector [83].

7.3.1 Harris Corner Detector

The Harris corner detector [56] is used to locate corners that exist within an image itself. The basic idea of the Harris corner detector is that a corner point should have significant changes in all the directions. Hence, the detector will compute the local

Fig. 7.6 The Gaussian
scale-space representation

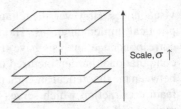

Scale, σ ↑

gradient of the neighbouring pixels that surround the candidate point. If the local
gradient is larger than a predefined threshold, the candidate point will be considered
as the feature point. The Harris corner detector is a type of fast and simple detector.
It is also invariant to noise, rotation and changes in illumination. In other words,
this detector is capable of locating the same feature point, even though the image
has been rotated or changed in illumination.

7.3.2 Harris–Laplace Detector

The Harris–Laplace detector [92] is an extension from the Harris corner detector.
Other than rotation and changes in illumination, this detector is invariant to changes
in scale as well. The idea of the Harris–Laplace detector is to create a Gaussian
scale-space representation of the original image as shown in Fig. 7.6 and then locate
the corner at different scales.

Each level in the Gaussian scale-space representation is computed by convolving
the images with the Gaussian kernel as shown in Eq. (7.1). In this case, x and y
are the horizontal and vertical distance from the origin, respectively, and σ is the
Gaussian distribution (scale).

$$G(x, y, \sigma) = \frac{1}{2\pi\sigma}e^{-(x^2+y^2)/2\sigma} \qquad (7.1)$$

However, the use of the Gaussian scale-space representation can increase
computational and memory complexity, since it is now required to search for the
corner at different scales.

7.3.3 Laplacian-of-Gaussian (LoG) Detector

The idea of the LoG detector [79] is similar to the Harris–Laplace detector. Just that
the LoG detector is used for blob (local maxima or local minima) detection. The
scale-space representation adopted in the LoG detector is slightly different from the
Harris–Laplace detector. In this case, each level in the scale-space representation is

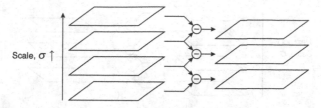

Fig. 7.7 The Difference-of-Gaussian (DoG)

taken as the difference between the two consecutive scales in the Gaussian scale-space representation as shown in Fig. 7.7

This scale-space representation is denoted as the difference of Gaussian (DoG). After the DoG is produced, the magnitude of a candidate point is compared to its eight neighbouring pixels that are located in the same scale plus the nine pixels in the next and the lower scales. If the candidate point appears to be the local maxima or minima, then it will be considered as the feature point. Since the feature points are to be located across the scales, the feature points generated from the LoG detector is also invariant to the changes in scale.

7.3.4   Features from Accelerated Segment Test (FAST) Detector

In the FAST detector [110], all the pixels that fall within the Bresenham's circle with a predefined radius from the candidate point will be tested against a predefined threshold. If a certain number of pixels are having the value that is larger or smaller than the predefined threshold, the candidate point will be considered as the feature point. The computation of this detector can be more efficient than the Harris corner detector since it involves mostly comparing the magnitude of all the pixels that fall within the Bresenham's circle.

7.3.5   Scale Invariant Feature Transform (SIFT)

The SIFT detector [83] has been adopted in many applications, due to the superior performance. As suggested by its name, this detector is invariant to scale changes as well. Moreover, it has been shown in [93] that the SIFT is also invariant to rotation, illumination and certain degree of changes in viewpoint. In the SIFT detector, the DoG as shown in Fig. 7.8 is first computed. This is different from the DoG used by the LoG detector. In this case, the SIFT will compute the DoG at different resolutions that are denoted as octaves. This DoG improves the robustness of the SIFT detector in locating the feature points at different scales.

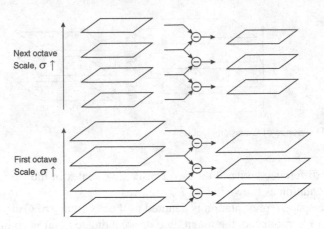

Fig. 7.8 The computation process of the DoG adopted in SIFT

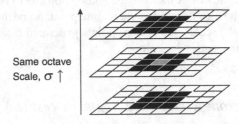

Fig. 7.9 Comparing the candidate point (*centre pixel, shaded*) to its neighbouring pixels (*darker surrounding pixels*)

Similar to the LoG detector, if a pixel is found to be the local maxima or minima after being compared to its eight neighbouring pixels plus nine pixels on the bottom and top level in the same octave as shown in Fig. 7.9, the pixel will be considered as the feature point. In this case, the pixel in grey colour is the candidate point, whereas the pixels in black colour are the neighbouring pixels.

7.3.6   Comparison of Feature Extraction Methods

A comparison of the pixel-based and the feature-based approaches is summarised in Table 7.1. The comparisons are made in terms of not only the computational and memory complexity but also the capability to locate the same feature points when there are changes in scale, rotation, illumination and viewpoint.

It can be seen that the capability of the SIFT detector is much better than the others since it is invariant to most of the changes. However, the memory usage of the SIFT detector is much higher than the others because it requires to construct the DoG at different scales and octaves. Although the Harris–Laplace detector and the

Table 7.1 Comparison of different approaches used for feature extraction

Detector	Memory usage	Complexity	Scale	Rotation	Illumination	Viewpoint
				Invariant		
Pixel based	Low	Moderate	X	X	△	X
	/	↓	/			
	Moderate*	Low#	O*			
Harris corner	Low	Low	X	O	O	X
Harris–Laplace	Moderate	Moderate	O	O	O	X
LoG	Moderate	Moderate	O	O	O	X
FAST	Low	Low	X	O	O	X
SIFT	High	Moderate	O	O	O	△

*Depends on whether coarse-to-fine searching is adopted
#Complexity reduces from the coarse layer to the fine layer due to less number of pixels

LoG detector also need to construct the scale-space representation of the original image, the number of layers that have to be constructed is less than the one adopted by the SIFT detector. The computational and memory complexity of the Harris corner detector and the FAST detector are the lowest among the detectors that have been described above. In the next section, the usage of feature extraction in image stitching, together with other additional processes required to complete the stitching, will be presented. It is important to note that the output from feature extraction is just a set of coordinates that indicate the location of the feature points. These feature points still need to be identified and matched to determine the overlapping regions.

7.4   Overview of Image Stitching

Generally, the stitching process can be divided into a few stages as shown in Fig. 7.10. For simplicity, it is assumed that only two images, which are denoted as images A and B, are to be stitched together. Before the stitching process is carried out, it is required to determine the dominant and non-dominant image. The non-dominant image will be transformed into the coordinate system that has been agreed by the dominant image. After the selection is done, the overlapping regions of the two images will be determined. This is achieved by searching for the similar feature points in the two images.

The output from feature extraction is just a set of coordinates that indicate where the feature points are located. Hence, it is necessary to "label" these feature points by using the feature point descriptor before they can be matched to each other. Once the matching is done, the transformation parameters are estimated by using the Random Sample Consensus (RANSAC) [42] algorithm and image stitching

Fig. 7.10 The main steps in
the image stitching process

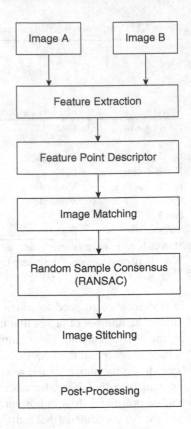

can be directly performed after this. Finally, some post-processing techniques such
as image compression or image blending can be applied to the stitched image.
The former is used to further compress the stitched image whereas the latter is
adopted to improve the visual quality of the stitched image. In the following
subsections, the stitching process will be discussed in more detail.

7.4.1   Feature Points Descriptor and Image Matching

After the feature points have been extracted from the images, it is necessary to
match them. The "identity" of a feature point is usually generated by referring
to the pixels around the feature point such as the local intensity or local gradient.
The descriptor that is used in the stitching process of the proposed frameworks
is the one adopted by the SIFT detector. In this case, the magnitude and orientation
of a feature point are first determined by using the local gradient. This will help
to improve the accuracy in finding the matched point if there is rotation. Then,
the magnitudes and orientations of the neighbouring pixels will be computed as
well. The sum of magnitudes and orientations will serve as the "identity" for the

Table 7.2 Different types of transformations and their respective transformation parameters

Mapping	Transformation parameters	Eq. no.	Illustration
Translation	$$\begin{bmatrix} x' \\ y' \end{bmatrix} = \begin{bmatrix} 1 & 0 & t_0 \\ 0 & 1 & t_1 \end{bmatrix} \begin{bmatrix} x \\ y \\ 1 \end{bmatrix}$$	(7.2)	
Rotation and translation	$$\begin{bmatrix} x' \\ y' \end{bmatrix} = \begin{bmatrix} \cos\theta & -\sin\theta & t_0 \\ \sin\theta & \cos\theta & t_1 \end{bmatrix} \begin{bmatrix} x \\ y \\ 1 \end{bmatrix}$$ θ = degree of rotation	(7.3)	
Scaled rotation and translation	$$\begin{bmatrix} x' \\ y' \end{bmatrix} = \begin{bmatrix} s(\cos\theta) & s(-\sin\theta) & t_0 \\ s(\sin\theta) & s(\cos\theta) & t_1 \end{bmatrix} \begin{bmatrix} x \\ y \\ 1 \end{bmatrix}$$ s = scale factor	(7.4)	
Affine	$$\begin{bmatrix} x' \\ y' \end{bmatrix} = \begin{bmatrix} t_0 & t_1 & t_2 \\ t_3 & t_4 & t_5 \end{bmatrix} \begin{bmatrix} x \\ y \\ 1 \end{bmatrix}$$	(7.5)	
Projective	$$\begin{bmatrix} \tilde{x} \\ \tilde{y} \\ \tilde{z} \end{bmatrix} = \begin{bmatrix} t_0 & t_1 & t_2 \\ t_3 & t_4 & t_5 \\ t_6 & t_7 & 1 \end{bmatrix} \begin{bmatrix} x \\ y \\ 1 \end{bmatrix}$$	(7.6)	
	$$x' = \frac{\tilde{x}}{\tilde{z}} = \frac{t_0 x + t_1 y + t_2}{t_6 x + t_7 y + 1}$$	(7.7)	
	$$y' = \frac{\tilde{y}}{\tilde{z}} = \frac{t_3 x + t_4 y + t_5}{t_6 x + t_7 y + 1}$$	(7.8)	

feature point when performing the matching. After the "identity" is generated, the matching of the feature points between the two images can be carried out. Some of the methods that can be used to match the feature points include exhaustive search and k-d tree.

7.4.2 Transformation Parameters and Image Stitching

After matching is completed, the next step is to transform the non-dominant image into the new coordinate system that has been agreed by the dominant image. Before the transformation can be performed, it is required to first determine the transformation parameters that decide how the pixels in an image are to be remapped into the new coordinate system. The number and the dimension of the transformation parameters are strongly related to the type of transformation. According to [119], the type of transformation can be divided into five types as shown in Table 7.2.

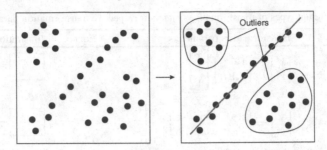

Fig. 7.11 Estimating a line within a set of data using RANSAC. All the data that is not fitted to the model is called the outliers

From the type of transformation shown in Table 7.2, it can be seen that the transformation parameters are different for each type of transformation. The complexity of the transformation parameters is also directly related to the difficulty of the transformation itself. On the one hand, only two parameters need to be determined if the transformation only consists of simple translation that involves movement in the vertical and horizontal direction. On the other hand, eight parameters are required to be estimated in the case of a projective transformation because the transformation involves not only translation but also rotation, scaling, shearing, etc. Generally, all these parameters are to be computed using the RANSAC [42] algorithm after a set of matched feature points is obtained. It is an iterative method used to estimate a model that is best fitted to a set of data. For example, it can be used to search for data that is fitted to a straight line model as shown in Fig. 7.11. Data that are not fitted to the model are denoted as the outliers, whereas the data fitted to the model are called the inliers.

This algorithm first starts by randomly choosing a subset of the complete data and consider these data as the inliers. Then, a temporary model that is fitted to these inliers is defined. Next, how well each of the remaining data is fitted to this model will be tested. If the error is smaller than the predefined threshold, the data is considered part of the model, and it will be added into the subset. When the number of data within the subset is large enough, it can be said that a good model has been found. All these steps will be repeated until the best model is found.

7.4.3 Post-Processing

Due to the automated exposure and aperture control of the image sensors, the images taken in the same place with two image sensors can have some obvious changes in terms of the intensity. Depending on the setup of the image sensors, the object that appears in the images can be slightly tilted. All these can lead to undesired effects to appear in the stitched image such as ghosting, visible seam and sharp transition of intensity from one side to another. Many techniques have been proposed to remove or to minimise all these undesired effects such as the weighted average [17], selective image composition [77], optimal seam selection [94] and multiband blending [15]. In the next section, the performance of stitching the subsampled images by using different feature extraction methods is investigated and presented.

7.5   Analysis of Feature Extraction Methods on Subsampled Images

The performance of feature detectors in extracting feature points from the subsampled versions of the original image is evaluated. This is useful if feature extraction is to be performed at the visual node directly. In this case, the minimum resolution and number of feature points that allow the stitching to be completed without any errors are investigated. The reason of using the subsampled images is to reduce the local memory requirement and the processing time for hardware implementation. Consider the case where a full resolution of an image is in the size of 640×480. In this case, 307,200 pixels are required to be processed. If the image is to be subsampled with a ratio of 2:1, the resolution will reduce to 320×240, and only 76,800 pixels are required to be processed. It will be easier to process the subsampled image than the complete image due to the lower resolution. In this section, the five types of feature-based approaches listed in Table 7.1 are evaluated by using the four set of images shown in Table 7.3. However, it is important to note that the subsampled images are only being used to estimate the transformation parameters. After the parameters are known, it will be used to transform and stitch the images in full resolution.

7.5.1   Minimum Resolution of Subsampled Images

The original images will be subsampled with respect to the ratio of 2:1, 4:1 and 8:1. The subsampling can be completed without interpolation which is more favourable for hardware implementation. The quality of the stitching is categorised into three groups as shown in Table 7.4, and the simulation results are summarised in Table 7.5. From the simulation results shown in Table 7.5, it can be seen that the best performance is achieved by the LoG detector. When the resolution of the original images is reduced to 160×120, the stitching can be completed without any errors. The performance of the Harris corner, Harris–Laplace and SIFT detectors is roughly the same. These detectors can successfully stitch the images when the resolution is reduced to 320×240, except for the FAST detector that experienced some problems in stitching the images from Sets 2 and 4. The number of feature points is strongly related to the success rate of stitching.

Table 7.3 The four sets of images that will be used to evaluate the detectors for image stitching. All the images are in greyscale with the original resolution of 640 × 480

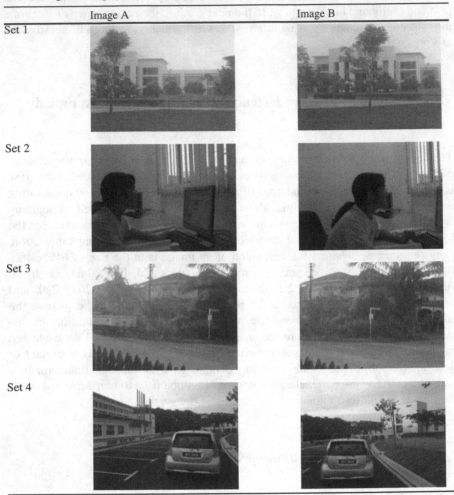

7.5.2   *Number of Feature Points*

The number of feature points extracted by using the Harris corner (Harris), Harris–Laplace (Laplace), LoG, FAST and SIFT detectors at different subsampled resolution is summarised in Fig. 7.12. By referring to the simulation results, it can be seen that when there is a substantial amount of feature points, the chances to successfully stitch the images is also higher. In the case of LoG detector, the number of feature points extracted from the subsampled image is by default being fixed to 120 [47]. This number is much higher than the number of feature points generated by other detectors. Hence, only LoG detector can complete the stitching when the

Table 7.4 The sample stitched image to define the stitching quality

Stitching without error	Stitching with minor errors	Fail

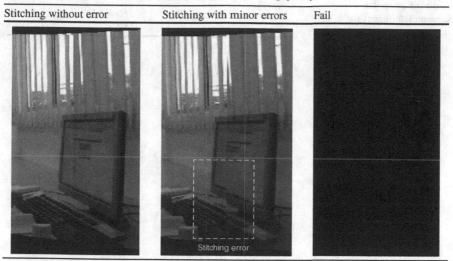

subsampled image with a resolution of 160×120 is adopted. In contrast, the number of feature points extracted from the FAST detector is much fewer than the other detectors. The FAST detector is only able to detect larger number of feature points in the images from Set 3.

When the number of feature points increases, a better model can be estimated by using the RANSAC algorithm, and the errors that may occur in the stitched image can be minimised. However, this will also increase the processing time since the number of matching operations will also increase significantly. When the number of feature points is close to 60, the chances to stitch the images successfully are sufficiently high. In order to show this, the sensitivity of the Harris corner detector has been increased so that more feature points will be extracted. Due to the increase in the number of feature points, it is now possible to complete the stitching by using subsampled images with the resolution of 160×120. The stitching of images from Set 2 to 4 can be completed without any errors except for the images from Set 1 that still produce minor errors. However, it still shows that increasing the number of feature points can help to increase the success rate of stitching. Overall, the success rate of stitching is still primarily determined by whether a substantial set of distinct feature points can be extracted from the subsampled images. However, it is not feasible to have a large amount of feature points as well, since it will increase the processing time due to the massive amount of matching operations. The number of feature points that can be extracted by a detector should be close to 60 or above to achieve a higher success rate in stitching. In addition, the resolution of 160×120 should be very close to the minimum resolution that still allows a substantial amount of feature points to be extracted out.

Table 7.5 The evaluation results of using the subsampled images for stitching

	Subsampled resolution		
	320 × 240 (2:1)	160 × 120 (4:1)	80 × 60 (8:1)
Harris corner detector			
Set 1	O	△	X
Set 2	O	△	X
Set 3	O	X	X
Set 4	O	△	X
Harris–Laplace detector			
Set 1	O	△	X
Set 2	O	△	X
Set 3	O	X	X
Set 4	O	△	X
LoG detector			
Set 1	O	O	X
Set 2	O	O	X
Set 3	O	O	X
Set 4	O	O	X
FAST detector			
Set 1	X	X	X
Set 2	△	X	X
Set 3	O	X	X
Set 4	O	△	X
SIFT detector			
Set 1	O	△	X
Set 2	O	△	X
Set 3	O	X	X
Set 4	O	△	X
Harris corner detector *(increased sensitivity)*			
Set 1	–	O	X
Set 2	–	O	X
Set 3	–	X	X
Set 4	–	O	X

Note:
O Stitching completed without any error
   (no disparity near the edges)
△ Stitching completed with minor errors
   (visible seam/disparity)
X Stitching cannot be completed
   (fail)

7.6 Proposed Framework

In the proposed framework, multiple vision nodes are used to capture the scene from different viewpoints. In this case, both the process of locating the overlapping regions and stitching will be performed at the aggregation node. Only a small set of tree pruning data which can help the visual node to discard the overlapping regions

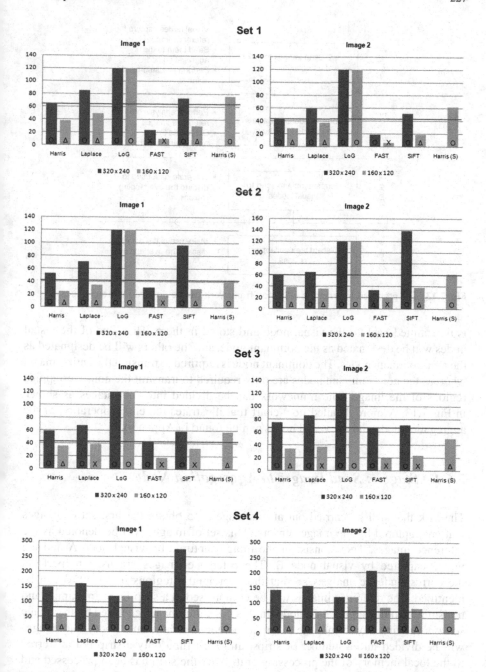

Fig. 7.12 Number of feature points extracted from Image A and B of Set 1–4

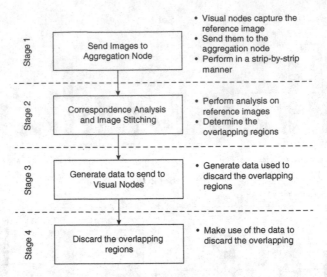

Fig. 7.13 The proposed information reduction framework can be divided into four stages

is transmitted back to the visual node and stored in there. Then one of the visual nodes will be designated as the dominant node, and the others will be designated as the non-dominant nodes. The dominant node is required to transmit the entire image, whereas the non-dominant nodes are only required to transmit the non-overlapping regions of the image. The framework can be divided into four stages as shown in Fig. 7.13. A more detailed flow chart that illustrates the entire operation of the aggregation node and the visual nodes can be found in Appendix I.

7.6.1 Stage 1: Send Images to Aggregation Node

The task that will be carried out at this stage is to obtain the first set of images that are captured by the image sensors. This set of images will be denoted as the reference images. They consist of the image captured by visual node A and the image captured by visual node B. The reference images are used to perform the correspondence analysis so that the overlapping regions in the FOV can be identified. The reference images will be compressed in a strip-by-strip manner and transmitted to the aggregation node. In this case, the images captured by the image sensors are first stored into the external memory on the visual node. The images will be divided into a number of strips, and each time only a strip is transferred to the local memory of the processing unit. After the strip has been processed and transmitted, another strip is loaded into the local memory again to continue the coding process. This is to reduce the local memory requirement. It is important to note that all the images captured by the visual nodes are to be compressed by using the one-pass downward scanning SPIHT coding. It is a modified version of the SPIHT coding presented in Chap. 6.

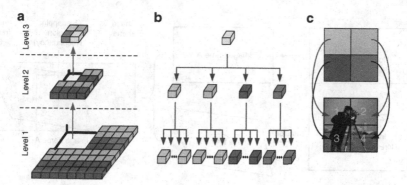

Fig. 7.14 (**a**) Distribution of the wavelet coefficients at different levels of decomposition. (**b**) SOT structure that links the wavelet coefficients shown in (**a**). (**c**) Regions that are covered by different sets of tree in the SOT structure

7.6.2   Stage 2: Correspondence Analysis and Image Stitching

After the reference images have been received by the aggregation node, the next step is to determine the overlapping regions. At this stage, the main task is to determine the overlapping regions through correspondence analysis (image stitching). The accuracy in locating the overlapping regions is strongly affected by the process of locating the feature points. Feature extraction plays an important role in the stitching process.

7.6.3   Stage 3: Generate Data for Visual Nodes

In the third stage, the transformation parameters obtained from the previous stage will be further processed. This is to create a new set of parameters so that the visual nodes can make use of these data to remove the overlapping regions easily. It is important to take note that all the processes described in this section are to be performed at the aggregation node. The tree pruning data are produced from analysing the SOT structure. After the overlapping regions are known, the wavelet coefficients that belong to the overlapping regions can be located. Then, the subtrees that link these wavelet coefficients together are determined, and this information will be used to produce the tree pruning data. The basic idea of tree pruning is to eliminate a set or a part of the tree in the SOT structure. This prevents the overlapping regions from being encoded by the SPIHT coding. As shown in Fig. 7.14, each tree structure corresponds to a specific region in the image. Therefore, discarding a tree structure will also remove the region covered by that particular tree structure. For example, if the rightmost tree structure shown in Fig. 7.14b is discarded, the region labelled with "4" in Fig. 7.14c will be removed.

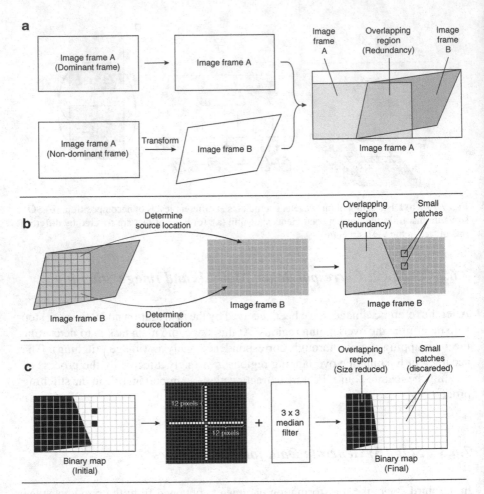

Fig. 7.15 (a) The process of identifying the overlapping region (redundancy). (b) The process of searching for the source location. (c) The process of extending the non-overlapping region to prevent the effect of aliasing

This stage involves several steps. First, it is necessary to create a binary map that decides the root of the tree structure that should be discarded before the SPIHT coding. The process of creating the binary map is illustrated in Fig. 7.15. This binary map is used to generate the tree pruning data. In this case, image A is considered as the dominant image, and image B is considered as the non-dominant image. After the overlapping regions are known, it is necessary to determine the source location where the pixels have to be selected to fill up the pixels in the overlapping regions. The reason for doing this is to identify the overlapping regions in the original image frame B. The source for the pixels that fall within the overlapping region in the original image frame B can be computed using Eqs. (7.7) and (7.8).

Algorithm 7.1 Non-overlapping region extension

---
      for all Pixels in the binary map **do**
2:       **if** Pixel value = 1(white) **then**
              12 pixels to the N, S, E, W = 1(logical)
4:       **end if**
      end for

The pixel located at coordinate (x', y') in the original image frame B is equivalent to the pixel located at coordinate (x, y) in the stitched image. Finally, a binary map is created to define the overlapping regions and the non-overlapping regions. The white colour region ("1") is used to indicate the pixels that are needed by the non-overlapping regions, and the black colour region ("0") is used to indicate the pixels that are needed by the overlapping regions. The pixels that fall within the black colour regions are considered to be the inter-redundancy that will be discarded.

A point to note is that the size of the overlapping region has to be reduced by extending the size of the non-overlapping region. This is to prevent aliasing to occur on those pixels that are situated near the borders between the overlapping and non-overlapping regions. This is due to the reason that part of the pixels in the overlapping region is actually involved in the process of calculating the DWT. Therefore, if all the pixels in the overlapping region are completely discarded, some information that is required by the inverse DWT will be lost. In this case, the process of extending the size of the non-overlapping region is achieved by using Algorithm 7.1. Generally, the idea is to ensure that for each pixel in the non-overlapping region, the 12 pixels located in the north, south, east and west direction of it are to be retained for the inverse DWT. When three levels of decomposition are used, this is the minimum number of pixels that is required to avoid from aliasing.

After the non-overlapping regions have been extended by using Algorithm 7.1, a 3×3-median filter is further applied onto the binary map. This is to remove all the small patches, which cannot be totally removed with Algorithm 7.1. The reason of doing this is to produce a smooth region that is more favourable for the SPIHT coding. These small patches can create a sharp transition in between the magnitude of the neighbouring pixels. When the DWT is applied, the sharp transition will increase the number of high-frequency components that affect the efficiency of the SPIHT coding as shown in Fig. 7.16. In this example, all the pixels in the test image are set to the value of 128 except for one of them in the middle is set to the value of 0. Then, the DWT with one level of decomposition is applied to the test image. This will be compared to the case where the value of all the pixels is set to 128.

The process carried out to obtain the tree pruning data is illustrated in Fig. 7.17. First, the binary map created from the previous section as shown in Fig. 7.17a will be used to produce another three binary maps which are generated for different levels of decomposition as depicted in Fig. 7.17b. It should be noted that the binary map for each level of decomposition is generated in a way that is similar to how the SOT structure is being constructed. In this case, a group of 2×2 pixels at the m level will contribute to a pixel at the $(m + 1)$ level as illustrated in Fig. 7.17b. If any of the

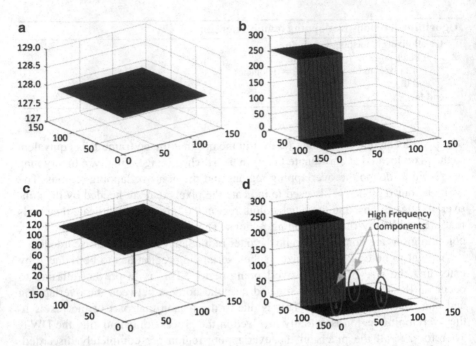

Fig. 7.16 (a) The test image with all the pixels value set to 128. (b) The outcome of (a) after one level of decomposition. (c) The test image with the middle pixel set to 0 and others remain at 128. (d) The high-frequency components created by applying one level of decomposition to (c)

pixels in the group carry a value of "1", the pixel at the upper level will be assigned with a value of "1". The pixel at the upper level is only assigned with a value of "0" when the entire group of 2×2 pixels have the value of "0".

The number of bits required to store the tree pruning data is equivalent to $[(W \times H) \times n_S] / 4$, whereby n_S is the number of strips and W and H are the width and height of the strip, respectively. If the original size of an image is 512×512, and it will be divided into 16 strips; the amount of memory needed to store the tree pruning data can be calculated as follows:

$$\text{Memory needed} = \frac{[(512 \times 32) \times 16]}{4} \text{ bits}$$

$$= 65536 \text{ bits}$$

$$= 8.192 \text{ kB}$$

Although the processing will be carried out in a strip-by-strip manner, 8.192 kB is the total amount of memory needed to store the tree pruning data for the entire frame. When the SPIHT coding is applied, it can just skip over the tree where the root falls in the black colour region defined in Fig. 7.17d. The pseudo code that describes the operation of the aggregation node is shown in Algorithm 7.2.

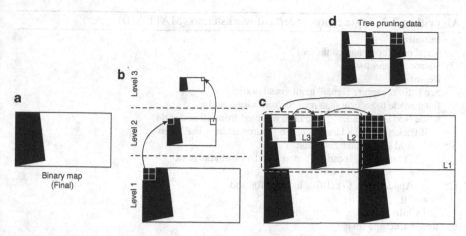

Fig. 7.17 (**a**) The binary map used to generate the tree pruning data. (**b**) The binary map created for different levels of decomposition. (**c**) The binary map in (**b**) is duplicated to cover all the sub-bands. (**d**) Each row of data in the *DESC* table is multiplied by the corresponding tree pruning data to obtain the new *DESC* table. After the binary map for levels 1 to 3 is produced, the binary map for each level is duplicated as shown in (**c**) in order to form a map that is fully correspondent to all the sub-bands. In this case, it is assumed that only three levels of decomposition are used. The tree pruning data is constructed by all the level 2 and level 3 sub-bands shown in (**c**). The level 1 sub-band is not included because it contains only the leave nodes. Based on the tree pruning data as shown in (**d**), if the root of a tree falls within the black colour region, then the entire tree is not required to be encoded by the SPIHT coding. The tree pruning data is best to use with the listless SPIHT coding

7.6.4 Stage 4: Discard the Overlapping Regions

At this stage, the main task is to use the new parameters generated from the previous stage to discard the overlapping regions. Visual nodes will use the mapping parameters and tree pruning data to discard the overlapping regions, respectively. In this case, the visual node that is designated as the dominant node will directly compress the image with SPIHT coding. Only the visual nodes that are designated as the non-dominant node will make use of the tree pruning data to discard the overlapping regions before the SPIHT coding is applied. After all the images are received by the host workstation, they will be stitched together after the SPIHT decoding. The pseudo code that describes the operation of the visual node is shown in Algorithm 7.3. In the next section, a one-pass downward scanning SPIHT coding that is designed to use with the tree pruning data will be described.

7.6.5 One-Pass Downward Scanning SPIHT Coding

Using the one-pass downward scanning, the SPIHT coding will directly encode each of the wavelet coefficients up to the predefined bit plane when performing

Algorithm 7.2 Aggregation node/host workstation (MATLAB)

    Initialize:
    Serial port (open and configure)
  3: Source address list
    Begin:
    Send "Start capture signal" to all visual nodes
  6: Enter mode to receive data packets from visual nodes
    while No "stop headers" have been received from *all* nodes **do**
      if Packet source address is *not* in "source address list" **then**
  9:     Add source address to list
        Create a new location to store the data from the node
      else
12:     Append data to existing location for node
      end if
    end while
15: Reconstruct the image
    if Images are reference types **then**
      Stitch them together
18:    Determine the overlapping regions
      Eliminate the small patches
      Generate the tree pruning data
21:    Broadcast the tree pruning data
    else
      Stitch them together
24: **end if**

    For next image frame
    goto: 4

---

the downward scanning. This helps to reduce the processing time since it is not necessary to scan through the wavelet coefficients for multiple times. The flow charts for the main program and the subroutines for the one-pass downward scanning SPIHT coding are shown in Figs. 7.18–7.21. The definitions and notations in the flow charts are explained as follows:

1. Memory address of the *SIG*, *DESC* and *GDESC* tables are initialised to 0.
2. The alphabet or number in the round bracket besides a term is used to represent the memory address where it points to. For example, *SIG*(5) refers to the coefficient that is located at memory address 5 in the *SIG* table.
3. The alphabets or numbers in the square bracket besides a term is used to indicate the bit plane number. For example, *DESC*(2) [3 : 5] refers to bit plane 3 to 5 of the coefficient that is located at memory address 2 in the *DESC* table.
4. Bit plane 1 is considered as the most significant bit (MSB).
5. The term "FOR" is used to indicate a for loop. After the term "FOR", the first field is used to initialise the variable, the second field is to specify the condition to terminate the loop and the third field is the increment of the variable. If the initialisation does not specify to any number, it means that it will be initialised based on the process that called the subroutine.

Algorithm 7.3 Visual node (assembly)

   Initialize:
   Clock rate
3: Resolution of the image
   Wait for broadcast signal from the aggregation node. {What signal? Is this the same "send data" signal from the node?}
   if Broadcast signal received **then**
6:    **Begin:**
      for n number of strips **do**
         if image $\neq$ reference image **then**
9:          Use the tree-pruning data to discard the overlapping regions
            Apply a high pass filter to the image pixels
            Apply a low pass filter to the image pixels. Store pixels into appropriate location. {what are these "appropriate locations"?}
12:         If desired level of decomposition is not achieved, repeat from Step 1. {I'm fuzzy about what Step 1 is here. Does it mean all the way from "init the clock rate...etc."?}
            Construct the SIG_TABLE, SIGN_TABLE, DESC_TABLE, GDESC_TABLE
            Perform the one-pass downward scanning listless SPHIT coding.
15:         Packetize the output bit stream and transmit to the aggregation node.
         else if image $=$ reference image **then**
            Wait for the configuration signal from the aggregation node
18:         Configure the node, dominant or non-dominant
         end if
      end for

21:   Repeat from line 4 to transmit another image
   end if

---

6. The subroutine used to check or to output the *DESC* and *GDESC* bits are the same. Hence, the operation that will be performed by the subroutine is determined by the process that called the subroutine.

7. If a subroutine is called, the memory address and the initialisation value specified by the process will be passed to the subroutine. For example, when the process CHECK *DESC*(z) BITS, INITIALIZE $d = 1$ calls the subroutine CHECK/OUTPUT *DESC*(b) BITS, the variable b will take the value of z and variable d will take the value of 1. Moreover, the check operation is performed instead of the output operation in this case.

Initially, it is necessary to initialise the number of bit planes to be encoded and set the pointers to point to the *SIG*, *DESC* and *GDESC* tables. In this case, y is pointed to the wavelet coefficients in the *SIG* table, and z is pointed to both the *DESC* and *GDESC* tables. Since the first 16 coefficients in the SIG table belong to the LL3 sub-band that does not have any offspring, it is only necessary to send the *SIG* bits and the *SIGN* bit when y is less than 16. This is also the reason why z is started at memory address 16 in the *DESC* and *GDESC* tables.

After the 16 coefficients in the LL3 sub-band are encoded, it is the turn for the coefficients that fall in the LH3, HL3 and HH3 sub-bands. These coefficients are located at memory address 16 to 63 in the *SIG* table. The *DESC* bits and *GDESC* bits for these coefficients will be located at the same memory address in the *DESC* and

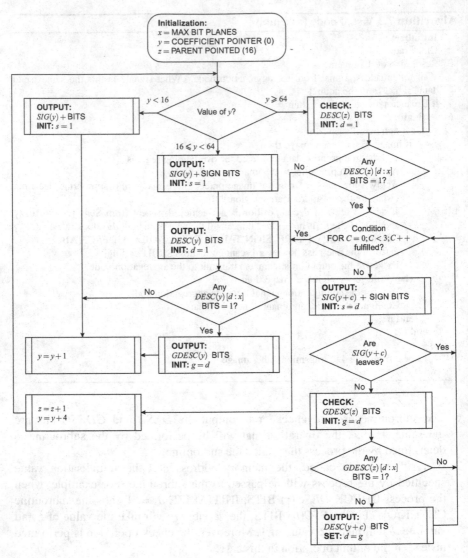

Fig. 7.18 The flow chart for main program of the one-pass downward scanning SPIHT coding

GDESC tables, respectively. Since the coefficients in these sub-bands are serving as the roots of the tree, it is also required to send the *DESC* bits and *GDESC* bits in addition to the *SIG* bits and *SIGN* bit. However, it should be noted that it is only necessary to send the *GDESC* bits when one of the bit planes in *DESC*(*y*) contains a "1". If that is the case, then the *GDESC* bits starting from the bit plane where *DESC*(*y*) is "1" have to be transmitted. Due to this reason, the location where the "1" is located in *DESC*(*y*) has to be saved prior to encoding the *GDESC*(*y*).

Fig. 7.19 Flow charts of the
subroutines for the main
program of the one-pass
downward scanning SPIHT
coding

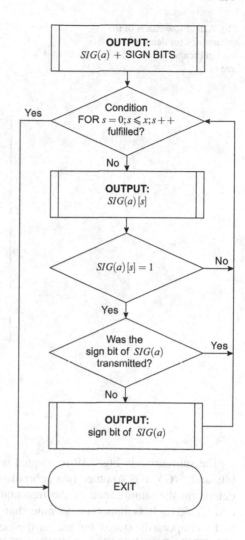

The process of encoding the remaining coefficients in the LH2, HL2, HH2, LH1, HL1 and HH1 sub-bands is identical. These coefficients are the offspring of the tree, and they are located at the memory address beyond 64 in the SIG table. At this stage, the coefficients in the SIG table are encoded in a group of 2×2. First, it is required to check the significance of the tree from $DESC(z)$. If one of the bit planes in $DESC(z)$ is "1", it implies that at least one of the offspring belonging to this tree is significant. In this case, the four direct offspring at $SIG(y + c)$ will be first encoded by sending the SIG bits and $SIGN$ bit. If the four direct offspring is not the leaves node, then it is required to check $GDESC(z)$ as well. This is to determine whether any of the offspring below the four direct offspring are significant. If the case is true, it will be reflected in one of the bit plane in $GDESC(z)$ with a "1", and the $DESC$ bits from $DESC(y + c)$ have to be transmitted.

Fig. 7.20 Flow charts of the subroutines for determining the significance of the *DESC* tree

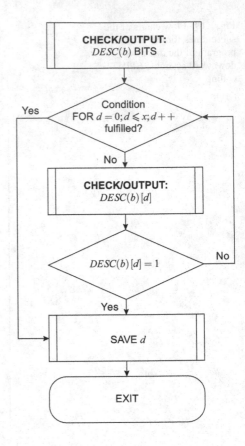

The subroutine in Fig. 7.19 is adopted when it is necessary to transmit the *SIG* bits and *SIGN* bit. Another two subroutines in Figs. 7.20 and 7.21 are used to determine the significance of the tree and output the respective bits for *DESC* and *GDESC*. It is important to note that the bit plane variable d and g need to be temporarily stored for use in the next stage of the processing. Since the SIG_PREV, DESC_PREV and GDESC_PREV tables are not required to keep track of the coding process, the overall memory required to perform the SPIHT coding is $[(W \times H) + (W \times H)/4 + (W \times H)/16]$ bits fewer than the previous implementation. In this case, W and H represent the width and the height of the full-size image, respectively.

It can be seen that the descendants are to be encoded based on the information provided by the DESC table. Hence, the operation of tree pruning can be easily carried out by modifying the value in the DESC table. After the roots of the trees that ought to be discarded have been located using the method shown in Fig. 7.17, these information are used to discard the trees by setting the value of the roots in the *DESC* table to "0". In this case, the SPIHT coding will just skip through the trees without having to encode them. This is another reason that this listless implementation is used.

Fig. 7.21 Flow charts of the
subroutines for determining
the significance of the
GDESC tree

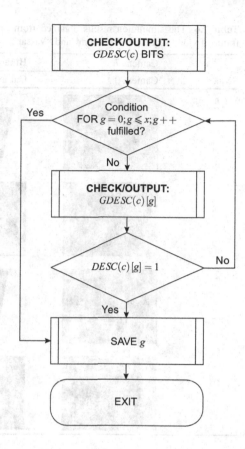

7.7 Simulation Results

Simulations were performed to evaluate the relationship between the percentage
of the overlapping region and the reduction in the number of bits. The proposed
framework is applied to the first frame of the video sequences, "Ballroom" and
"Vassar". These two video sequences are stereoscopic video sequences obtained
from [90]. In this case, the sample frames captured by the two cameras located at
the left most (camera 0) and the right most (camera 7) are used. The two cameras
are separated by 20 cm apart from each other. The sample frames are subsampled to
reduce the frame size from 640×480 to 128×96. This is closer to the frame size that
will be adopted for the hardware implementation. The sample frame from camera 7
is selected as the dominant frame, and the sample frame from camera 0 is selected as
the non-dominant frame. For feature extraction, the LoG detector is adopted. After
the overlapping region in the sample frame from camera 0 is identified, the binary
map will be produced to generate the tree pruning data. The results are summarised
in Table 7.6. It can be seen that the region that is discarded is slightly different
from the binary map. This is due to the nature of the SOT structure that each of the

Table 7.6 The simulation results obtained from applying the proposed framework to the first frame of video sequence "Ballroom" and "Vassar"

Sequence	Camera 7 Camera 0	Binary map (Final) Camera 0 (Final)	Stitched image
Set 1			
Set 2			

nodes has either no offspring or four offspring. In other words, only two options are available when pruning the tree which is either to discard nothing or to remove the four offspring at one time.

7.7.1   Relationship Between Overlapping Regions and Number of Output Bits

The original sample frame and the non-overlapping region of the sample frame from camera 0 were encoded using the SPIHT coding. The size of the resulting bit stream is summarised in Table 7.7. From the simulation results, it can be seen that the proposed framework can significantly reduce the number of bits that need to be transmitted by the visual node. However, the actual number of wavelet coefficients that can be discarded is smaller than what is defined by the final binary map. This is to ensure that the pixels falling near the edges between the overlapping and non-overlapping regions can be reconstructed perfectly. If a small distortion can

Table 7.7 The percentage of overlap (%) and the reduction in number of bits that are required to transmit the non-overlapping region

Sequence	Overlap (%)	Original (number of bits)	Non-overlapping region (number of bits)	Bits reduction (%)
Ballroom (Frame 1)	89.56	82,865	35,227	57.49
Vassar (Frame 1)	85.99	74,318	36,460	50.94

be tolerated, then the bits reduction (%) should be able to get closer to the actual overlap (%).

7.8 Hardware Implementation

In order to accommodate the process of using the tree pruning data to modify the DESC table, the strip-based MIPS architecture described in Chap. 4 has been modified. The modified architecture together with the controller is illustrated in Fig. 7.22. The tasks that can be handled by the controller are listed as follows:

1. Control the frame capture process.
2. Control the transferring of image data between the external memory and data memory.
3. Initiate and control the transmission of output data.
4. Initiate and control the receiving of tree pruning data.
5. Discard the tree based on the tree pruning data.

The tree pruning data are directly stored in the local memory so that they can be easily accessed by the processor. The amount of memory needed is calculated as $[(W \times H) \times n_S]/4$, where W represents the width of a strip, H represents the height of a strip and n_S represents the number of strips. In this case, the value of W, H and n_S is 128, 8 and 16, respectively. Hence, the total amount of local memory needed is 512 B. The controller will explicitly modify the $DESC$ table located in the data memory with respect to the tree pruning data. After the root of the tree that ought to be discarded has been located in the $DESC$ table, this tree can be easily removed by setting the value of the root to zero. In this case, the SPIHT coding will not encode any coefficients that belong to this tree.

For the SPIHT coding, the DWT is performed by using the LeGall 5/3 implementation. In order to reduce the memory usage, the strip size was reduced from 16×128 to 8×128. This not only reduces the memory requirement that is imposed on the data memory by 50 % but also the temporary buffer that is used to store the rearranged DWT coefficients. Since this changes the strip size, it is necessary to re-compute the address calculations. In this case, the new address calculations that are generated to work with the strip size of 8×128 are summarised in Table 7.8.

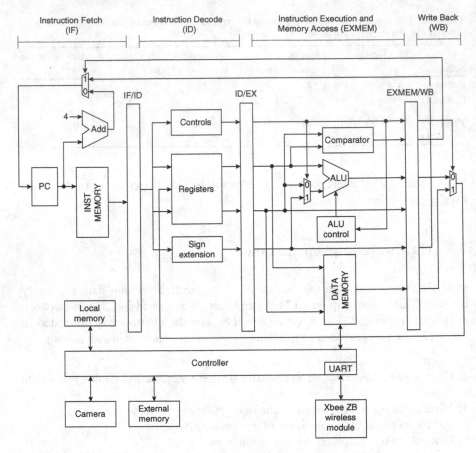

Fig. 7.22 Modified MIPS architecture to accommodate the use of tree pruning

Table 7.8 The predefined rules used to calculate the new address where the wavelet coefficients are positioned in the *SIG* table. In this case, A_9 is the MSB and A_0 is the LSB

l-level DWT decomposition	$i=1$ (MSB)	(LSB)	$i=2$ (MSB)	(LSB)	$i=3$ (MSB)	(LSB)
Initial image pixel address	$A_9A_8A_7A_6A_5A_4A_3A_2A_1A_0$		–		–	
Initial LL pixel address	–		$A_7A_6A_5A_4A_3A_2A_1A_0$		$A_5A_4A_3A_2A_1A_0$	
New address of DWT coefficient in STRIP_RAM	$A_7/A_0 =$ 1(HL, LH, HH) : $A_7A_0A_6A_5A_4A_3A_2A_9A_8A_1$ else (LL): $A_7A_0A_9A_8A_6A_5A_4A_3A_2A_1$		$A_6/A_0 =$ 1(HL, LH, HH) : $A_6A_0A_5A_4A_3A_2A_7A_1$ else (LL): $A_6A_0A_7A_5A_4A_3A_2A_1$		$A_5A_0A_4A_3A_2A_1$ $A_5A_0A_4A_3A_2A_1$	

Table 7.9   Number of MIPS instructions used to process the image data

Process	Number of MIPS instructions
5/3 discrete wavelet transform + rearranging coefficients	230
Pre-collect Information for *SIG*, *DESC* and *GDESC* tables	131
One-pass downward scanning SPIHT coding	400

Table 7.10   Total number of clock cycles required to encode all the 16 strips of an image frame

Process	Clock cycles	Proportion (%)
Load data from external memory to local memory	81,952	1.7
5/3 discrete wavelet transform + rearranging coefficients	1,687,168	36.3
Pre-collect Information for *SIG*, *DESC* and *GDESC* tables	629,408	13.5
Tree pruning	17,440	0.4
One-pass downward scanning SPIHT coding	2,158,592–2,297,856	48.1
Total	4,574,560–4,713,824	100.0

7.9   Experimental Results

The MIPS architecture presented in Fig. 7.22 is implemented as a soft-core processor in the Virtex-II FPGA on the development board. This soft-core processor is adopted to perform the DWT on the image data, pre-collect the information of the wavelet coefficients and store them into the *SIG*, *DESC* and *GDESC* tables and compress the image data with the SPIHT coding. The number of MIPS instructions used to perform these functions is summarised in Table 7.9.

The communication between the development board and the wireless module is handled explicitly with the external controller. As described previously, the external controller is used to control the other peripherals, such as the external memory and camera. The camera is configured to capture image by using the YC_bC_r colour space, and only the Y component is processed. All the data is stored into the external memory before they are processed by the soft-core processor. Currently, the visual node is set to run at the clock speed of 25.175 MHz. The total number of clock cycles required to process all the 16 strips of an image frame is summarised in Table 7.10.

The SPIHT coding is configured to encode up to the 8th bit plane. The number of clock cycles occupied by the SPIHT coding will vary according to the content of an image frame. In this case, the visual node is capable to process at the speed of 5.3–5.5 frames per second (fps). The amount of resources occupied by the soft-core processor, together with the external controller and the other peripherals, is summarised in Table 7.11.

The utilisation where the tree pruning function is completely removed from the controller is also investigated. From the results, it can be seen that there is only a minor increase in the total utilisation when tree pruning is adopted. However, the number of clock cycles required to process the entire image and the amount of utilisation can be further reduced after optimisation. The setup of the proposed

Table 7.11 Resources utilisation of the soft-core processor

	Utilisation	
Resources	With tree pruning	Without tree pruning
Occupied slices	4,510/14,336 (31 %)	4,220/14,336 (29 %)
Slice flip-flops	2,915/28,672 (10 %)	2,624/28,672 (9 %)
4-input lookup table (LUT)	6,714/28,672 (23 %)	6,318/28,672 (22 %)
16-bit block RAM	28/96 (29 %)	27/96 (28 %)

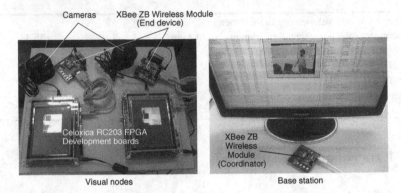

Fig. 7.23 Hardware experiment setup

system is shown in Fig. 7.23. The host workstation is a PC connected to the wireless module. The wireless module connected to the development board and the PC is identical. However, it should be noted that the module is configured to serve as the end device in the former case, whereas another module is configured to function as the base station (coordinator) in the latter case. In addition, all the wireless modules are operated at the default baud rate of 9600, and the maximum payload size which is 84 bytes is selected.

The image frames received from the visual nodes will be processed inside the PC using MATLAB. In this experiment, frames from visual node A are selected to be the dominant frame. Sample image frames captured by visual node A, B and the resulting stitched image are shown in Fig. 7.24. The first frame captured by visual node A and B is reference frame. These are the image frames that are going to be used to perform the stitching process. The binary map used to generate the tree pruning data is shown in Fig. 7.25. It can be observed from Fig. 7.25b that some small patches (outlined), still exist within the binary map after Algorithm 7.2 are applied. These small patches have to be removed with the 3×3 median filter.

After the stitching process has been completed, the parameters used to perform the stitching will be stored in the PC, and they will be reused to stitch the incoming image frames. From the simulation results, it can be seen that the distortion on the new object that is moving around the seam is almost negligible. This implies that the appearance of new objects does not create significant geometry changes. Therefore, the parameters can be reused for incoming new images without having to perform the stitching process all over again.

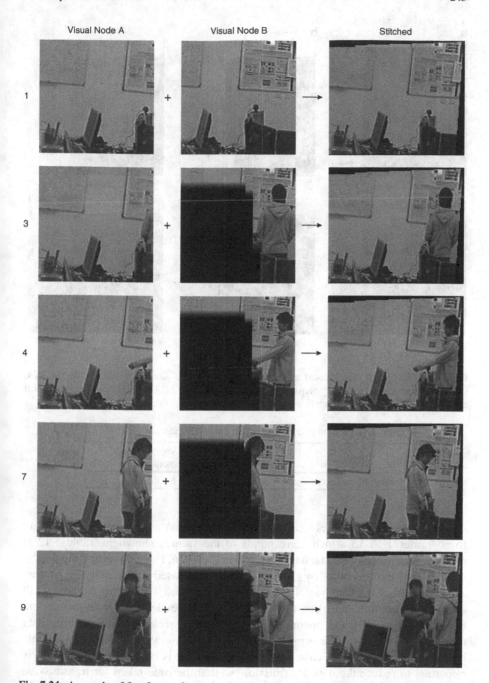

Fig. 7.24 A sample of five frames from visual node A, B and the resultant stitch

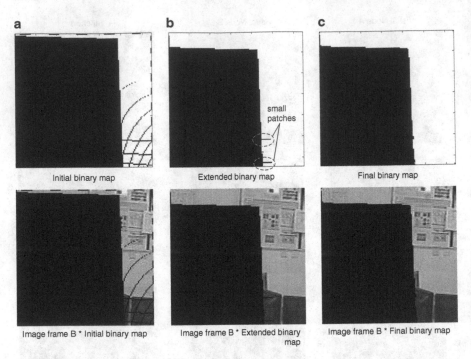

Fig. 7.25 The binary map created at different stages for the simulation shown in Fig. 7.24. (**a**) Initial binary map. (**b**) Binary map extended using Algorithm 7.2. (**c**) Final binary map obtained after the 3 × 3 median filter is applied

Table 7.12 The number of cycles that the battery can sustain

	Visual node A (dominant node)	Visual node B (nondominant node)
Number of cycles (images)	1,081	1,805

The total number of bytes received from visual node A and visual node B starting from frames 1 to 15 which corresponds to the frames shown in Table 7.12 is recorded by the host workstation and shown in Fig. 7.26. From the graph shown, the number of bytes transmitted by visual node B has decreased significantly from frame 2 onwards from the removal of the overlapping region. This reduces the overall energy consumption, since most of the energy is expended for data transmission. By referring to [65], the operating current of the wireless module during data transmission is 40 mA. This is much higher than the case when the wireless module is in the idle state. In this case, the operating current is only 15 mA. Hence, it is important to reduce the data transmission, so that the time taken for transmission can be lowered and lead to a decrease in the overall energy consumption.

As mentioned previously, the energy required to transmit the data is much higher than the energy needed to process the data. Hence, the focus is given to the energy consumed by the wireless module. In this evaluation, a brand new 9 V battery is

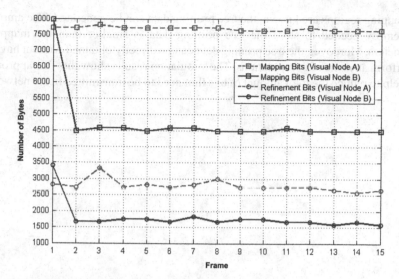

Fig. 7.26 The total number of mapping bits and refinement bits that are received from visual node A and visual node B for frames 1 to 15. The peak in frame 1 is because visual node B has to send the entire image frame to identify the overlapping region

used to power the wireless module, and the number of cycles it can sustain for the transmission of data in the proposed system is observed. The term number of cycles is referring to the number of images transmitted by visual node A (dominant node) and visual node B (non-dominant node). In addition, the duty cycle is set to 100 % (no sleep), and the visual nodes are located 5 m away from the coordinator. By default, the transmission power of the wireless module is +3 dBm. The results obtained from this evaluation are summarised in Table 7.12.

From the results, it can be seen that the lifetime of visual node B is longer than visual node A. After 1,081 cycles, the aggregation node can only receive the image from visual node B. This is due to the reason that the number of bytes required to be sent by visual node B is fewer than visual node A. From our observation, the wireless module is capable to transmit approximately 11 million bytes of data before the battery is depleted. Hence, compression is necessary to reduce the size of the images so that the lifetime of the visual node can be extended.

7.10 Chapter Summary

In this chapter, the proposed framework is designed for visual nodes that come with only one image sensor. Only one of the visual nodes which is denoted as the dominant node is responsible to transmit the entire image. Other visual nodes are only required to transmit the non-overlapping regions of the image. After the overlapping regions are known, the tree pruning data can be generated. The amount of memory needed to store the tree pruning data is smaller than the mapping parameters. Moreover, the tree pruning data can be directly obtained without having to perform any sort of computations. The simulation results show that this approach can help to reduce the total amount of bits that need to be transmitted in the network.

Appendix A

This section of the appendix will list the Handel-C code for the WMSN FPGA processor implementing the DWT.

```
// Handel-C Code for WMSN FPGA Processor Implementing the DWT

// === Pre-processor includes === //
#define PAL_TARGET_CLOCK_RATE 37500000
#include "pal_master.hch"
#include "stdlib.hch"

// === Parameter initializations === //
macro expr ClockRate         = PAL_ACTUAL_CLOCK_RATE;

macro expr ImageWidth        = 128;
macro expr ImageHeight       = 128;
macro expr StripHeight       = 16;
macro expr PatchWidth        = 8; //8;
macro expr PatchHeight       = 8; //8;
macro expr ImageBits         = 6;
macro expr StripBits         = 16;   // half of the standard 32-bits

macro expr NumStrips         = ImageHeight / StripHeight;
macro expr NumPatches        = (ImageWidth * StripHeight) / (PatchWidth *
    PatchHeight);

macro expr PatchesPerRow     = StripHeight / PatchHeight;
macro expr PatchesPerCol     = ImageWidth / PatchHeight;

macro expr RAMImageAddress   = ImageWidth * ImageHeight;
macro expr RAMStripAddress   = 8192;
macro expr InstMemSize       = 2048; // Allocated instruction memory size

macro expr RAMImageAddressBits = log2ceil(RAMImageAddress);
macro expr RAMStripAddressBits = log2ceil(RAMStripAddress);

macro expr StripCountBits      = log2ceil(NumStrips) + 1;
macro expr PatchCountBits      = log2ceil(NumPatches) + 1;

macro expr ImagePixelCountBits = RAMImageAddressBits + 1;
macro expr StripPixelCountBits = RAMStripAddressBits - 2 + 1;
macro expr PatchPixelCountBits = log2ceil(PatchWidth * StripHeight) + 1;
macro expr PatchWidthCountBits = log2ceil(PatchWidth) + 1;

macro proc ReadCamera();
macro proc LoadStripMemory();
macro proc RunProcessor();
```

L. Ang et al., *Wireless Multimedia Sensor Networks on Reconfigurable Hardware*, 249
DOI 10.1007/978-3-642-38203-1, © Springer-Verlag Berlin Heidelberg 2013

```
macro proc OutputStripMemory();

// Component declarations

ram unsigned int ImageBits IMAGE_MEMORY[RAMImageAddress] with {block = 1};
ram unsigned int StripBits STRIP_MEMORY[RAMStripAddress] with {block = 1};


unsigned int RAMImageAddressBits RAMAddressImageMemory;
unsigned int RAMStripAddressBits RAMAddressStripMemory;

unsigned int StripCountBits StripCounter;

void main (void)
{
    par
    {   RC10LEDWriteMask(15);                   // Light up four LSB LEDs to
        indicate start of processing
        // Run PAL/PSL functions
        RC10CameraRun(OV9650_RGB565_QVGA_LowLight,ClockRate/2);        //
            camera runs at 12MHz
        RC10RS232Run (RC10RS232_9600Baud,RC10RS232ParityNone,
            RC10RS232FlowControlHard,ClockRate);
        seq
        {
            ReadCamera();
            RAMAddressImageMemory = 0;
            StripCounter = 0;

            while (StripCounter != NumStrips)
            {
                LoadStripMemory();
                RunProcessor();
                OutputStripMemory();
                StripCounter++;
            }

            RC10LEDWriteMask(255);        // Light up remaining four LEDs to
                indicate all functions have been performed
        }
    }

} // end main

static macro proc ReadCamera()
{
    unsigned int CameraAddressX;
    unsigned int CameraAddressY;
    unsigned int CameraPixel;
    unsigned int ImagePixelCountBits ImagePixelCounter;
    unsigned int RAMImageAddressBits ImageAddress;

    seq
    {
        RAMAddressImageMemory = 0;
        ImagePixelCounter    = 0;
        while (ImagePixelCounter != RAMImageAddress)
        {
            par
            {
                IMAGE_MEMORY[RAMAddressImageMemory] = 0;
                RAMAddressImageMemory++;
                ImagePixelCounter++;
            }
        }
        do
        {
```

```
                    RC10CameraReadRGB565(&CameraAddressX,&CameraAddressY,&
                        CameraPixel);
            } while (!(CameraAddressX == 96 && CameraAddressY == 56));

            ImageAddress = 0;

            // Take the center portion of the 240x320 capture
            do
    {
      par
                {
                    RC10CameraReadRGB565(&CameraAddressX,&CameraAddressY,&
                        CameraPixel);
                    if (CameraAddressX <= 223 && CameraAddressX >= 96)
                    {
                        par
                        {
                            IMAGE_MEMORY[ImageAddress] = 0@CameraPixel[10:5];
                            ImageAddress++;
                        }
                    }
                    else
                    {
                        delay;
                    }
                }
    } while (CameraAddressY != 184);
            }
}

static macro proc LoadStripMemory()
{
    unsigned int StripPixelCountBits StripPixelCounter;
    unsigned int PatchCountBits      PatchCounter;
    unsigned int ImageBits           WriteData;
    seq
    {
        RAMAddressStripMemory = 0;
        StripPixelCounter    = 0;
        WriteData            = 0@IMAGE_MEMORY[RAMAddressImageMemory];
        RAMAddressImageMemory++;
        while (StripPixelCounter != (ImageWidth * StripHeight))
        {
            par
            {
                WriteData = IMAGE_MEMORY[RAMAddressImageMemory];
                STRIP_MEMORY[RAMAddressStripMemory] = 0@WriteData;
                RAMAddressImageMemory++;
                RAMAddressStripMemory++;
                StripPixelCounter++;
            }
        }
        // Initialize strip patches to valid
        RAMAddressImageMemory--;
        RAMAddressStripMemory = RAMStripAddress - NumPatches;
        PatchCounter = 0;
        while (PatchCounter != NumPatches)
        {
            par
            {
                STRIP_MEMORY[RAMAddressStripMemory] = 1;
                RAMAddressStripMemory++;
                PatchCounter++;
            }
        }
    }
}
```

```
static macro proc RunProcessor()
{
    // === Instruction Memory === //
    static ram unsigned int 32 INST_MEMORY[InstMemSize] =
    {
0b00100000000000110000000000000000, L0. ADDI $R0 $R3 0  \\ Start of
    Algorithm 4.11
0b00100000000000100000100000000000, L1. ADDI $R0 $R4 4096
0b00100000000000101000010000000000, L2. ADDI $R0 $R5 2048
0b00100000000000110000000000000000, L3. ADDI $R0 $R6 0
0b00000000110001011011000000101010, L4. SLT $R6 $R5 $R22 \\ while $R6 < $R5
    do
0b00010110110000010000000000001110, L5. BNE $R22 $R1 14 \\ branch to line 14
    if the condition above is not true
0b00000000000000000000000000000000, L6. SLL $R0 $R0 $R0
0b10001100011001110000000000000000, L7. LW $R3 $R7 0
0b00000000011000010001100000100000, L8. ADD $R3 $R1 $R3
0b10101100100001110000000000000000, L9. SW $R4 $R7 0
0b00000000011000010010000000100000, L10. ADD $R4 $R1 $R4
0b00000000011000010011000000100000, L11. ADD $R6 $R1 $R6
0b00010000000000000000000000000100, L12. BEQ $R0 $R0 4
0b00000000000000000000000000000000, L13. SLL $R0 $R0 $R0 \\ End of Algorithm
    4.11
0b00100000000000110000000000010000, L14. ADDI $R0 $R3 16 \\ Start of
    Algorithm 4.12
0b00100000000000100000000010000000, L15. ADDI $R0 $R4 128
0b00100000000000101000010000000000, L16. ADDI $R0 $R5 2048
0b00100000000000110000000000000001, L17. ADDI $R0 $R6 1
0b00100000000000111000100000000000, L18. ADDI $R0 $R7 4096
0b00100000000000100000010000000000, L19. ADDI $R0 $R8 4096
0b00000000000000010001000100100000, L20. ADD $R0 $R0 $R17
0b00000000000000010010000100100000, L21. ADD $R0 $R0 $R18
0b00000000000000010011000100100000, L22. ADD $R0 $R0 $R19
0b00000000000000010100000100100000, L23. ADD $R0 $R0 $R20
0b00000000000000010010000100100000, L24. ADD $R0 $R0 $R9
0b00000000000000010101000100100000, L25. ADD $R0 $R0 $R10
0b00000000010000101000000000100010, L26. SUB $R4 $R2 $R16
0b00000000100100111011000000101010, L27. SLT $R9 $R3 $R22 \\ while $R9 < $R3
    do
0b00010100001011000000000001010111, L28. BNE $R1 $R22 87  \\ branch to line
    87 if the condition above is not true
0b00000000000000000000000000000000, L29. SLL $R0 $R0 $R0
0b00000000101000100101100000101010, L30. SLT $R10 $R4 $R22 \\ while $R10 <
    $R4 do
0b00010110110000010000000001010011, L31. BNE $R22 $R1 83 \\ branch to line
    83 if the condition above is not true
0b00000000000000000000000000000000, L32. SLL $R0 $R0 $R0
0b00010101010101000000000000101001, L33. BNE $R10 $R16 41
0b00000000000000000000000000000000, L34. SLL $R0 $R0 $R0
0b10001100111100100000000000000000, L35. LW $R7 $R18 0  \\ Start of
    Algorithm 4.13
0b00000000011100010011100000100000, L36. ADD $R7 $R1 $R7
0b10001100111100110000000000000000, L37. LW $R7 $R19 0
0b00000000011100010011100000100000, L38. ADD $R7 $R1 $R7
0b00010000000000000000000000101110, L39. BEQ $R0 $R0 46
0b00000000000000000000000000000000, L40. SLL $R0 $R0 $R0
0b10001100111100100000000000000000, L41. LW $R7 $R18 0
0b00000000011100010011100000100000, L42. ADD $R7 $R1 $R7
0b10001100111100110000000000000000, L43. LW $R7 $R19 0
0b00000000011100010011100000100000, L44. ADD $R7 $R1 $R7
0b10001100111101000000000000000000, L45. LW $R7 $R20 0  \\ End of Algorithm
    4.13
0b00010101010100000000000000111011, L46. BNE $R10 $R0 59 \\ Start of
    Algorithm 4.14
0b00000000000000000000000000000000, L47. SLL $R0 $R0 $R0
0b00000010010101001010100000110000, L48. ADDS $R18 $R20 $R21
0b00000010011101011001100000100010, L49. SUB $R19 $R21 $R19
```

```
0b00000010011100111010100000110010, L50. ADDSS $R19 $R19 $R21
0b00000010101100101001000000100000, L51. ADD $R21 $R18 $R18
0b00000010011000010001000000100000, L52. ADD $R19 $R0 $R17
0b10101101000100100000000000000000, L53. SW $R8 $R18 0
0b00000001000000010100000000100000, L54. ADD $R8 $R1 $R8
0b10101101000100110000000000000000, L55. SW $R8 $R19 0
0b00000001000000010100000000100000, L56. ADD $R8 $R1 $R8
0b00010000000000000000000001010000, L57. BEQ $R0 $R0 80
0b00000000000000000000000000000000, L58. SLL $R0 $R0 $R0
0b00010101010100000000000001000111, L59. BNE $R10 $R16 71
0b00000000000000000000000000000000, L60. SLL $R0 $R0 $R0
0b00000010010100101010100000110000, L61. ADDS $R18 $R18 $R21
0b00000010011110101100110000100010, L62. SUB $R19 $R21 $R19
0b00000010001100111010100000110010, L63. ADDSS $R17 $R19 $R21
0b00000010101100101001000000100000, L64. ADD $R21 $R18 $R18
0b10101101000100100000000000000000, L65. SW $R8 $R18 0
0b00000001000000010100000000100000, L66. ADD $R8 $R1 $R8
0b10101101000100110000000000000000, L67. SW $R8 $R19 0
0b00000001000000010100000000100000, L68. ADD $R8 $R1 $R8
0b00010000000000000000000001010000, L69. BEQ $R0 $R0 80
0b00000000000000000000000000000000, L70. SLL $R0 $R0 $R0
0b00000010010101001010100000110000, L71. ADDS $R18 $R20 $R21
0b00000010011110101100110000100010, L72. SUB $R19 $R21 $R19
0b00000010001100111010100000110010, L73. ADDSS $R17 $R19 $R21
0b00000010101100101001000000100000, L74. ADD $R21 $R18 $R18
0b00000010011000010001000000100000, L75. ADD $R19 $R0 $R17
0b10101101000100100000000000000000, L76. SW $R8 $R18 0
0b00000001000000010100000000100000, L77. ADD $R8 $R1 $R8
0b10101101000100110000000000000000, L78. SW $R8 $R19 0
0b00000001000000010100000000100000, L79. ADD $R8 $R1 $R8 \\ End of Algorithm
    4.14
0b00000010100001001010000000100000, L80. ADD $R10 $R2 $R10
0b00010000000000000000000000011110, L81. BEQ $R0 $R0 30 \\ end while for
    $R10 < $R4 do
0b00000000000000000000000000000000, L82. SLL $R0 $R0 $R0
0b00000000000000000101000000100000, L83. ADD $R0 $R0 $R10
0b00000001001000010100100000100000, L84. ADD $R9 $R1 $R9
0b00010000000000000000000000011011, L85. BEQ $R0 $R0 27 \\ end while for $R9
    < $R3 do
0b00000000000000000000000000000000, L86. SLL $R0 $R0 $R0
0b00100000000001110001000000000000, L87. ADDI $R0 $R7 4096
0b00100000000001000001000000000000, L88. ADDI $R0 $R8 4096
0b00100000000011000001000000000000, L89. ADDI $R0 $R12 4096
0b00100000000110100000100000000000, L90. ADDI $R0 $R26 2048
0b00000000000000000100100000100000, L91. ADD $R0 $R0 $R9
0b00000000000000000101000000100000, L92. ADD $R0 $R0 $R10
0b00000000011000101000000000100010, L93. SUB $R3 $R2 $R16
0b00000000000000001000100000100000, L94. ADD $R0 $R0 $R17
0b00000000000000001001000000100000, L95. ADD $R0 $R0 $R18
0b00000000000000001001100000100000, L96. ADD $R0 $R0 $R19
0b00000000000000001010000000100000, L97. ADD $R0 $R0 $R20
0b00000010100010010110000000101010, L98. SLT $R10 $R4 $R22 \\ while $R10 <
    $R4 do
0b00010110110000010000000010111000, L99. BNE $R22 $R1 184 \\ branch to line
    184 if the condition above is not true.
0b00000000000000000000000000000000, L100. SLL $R0 $R0 $R0
0b00000010010001110110000000101010, L101. SLT $R9 $R3 $R22 \\ while $R9 <
    $R3 do
0b00010110110000010000000010110000, L102. BNE $R22 $R1 176 \\ branch to line
    176 if the condition above is not true.
0b00000000000000000000000000000000, L103. SLL $R0 $R0 $R0
0b00010101001100000000000001101111, L104. BNE $R9 $R16 111 \\ Start of
    Algorithm 4.15
0b00000000000000000000000000000000, L105. SLL $R0 $R0 $R0
0b10001100111100100000000000000000, L106. LW $R7 $R18 0
0b00000000111001000011100000100000, L107. ADD $R7 $R4 $R7
0b10001100111100110000000000000000, L108. LW $R7 $R19 0
```

```
0b0001000000000000000000000001110100, L109. BEQ $R0 $R0 116
0b00000000000000000000000000000000000, L110. SLL $R0 $R0 $R0
0b10001100111110010000000000000000000, L111. LW $R7 $R18 0
0b00000000011100100001110000000100000, L112. ADD $R7 $R4 $R7
0b10001100111110011000000000000000000, L113. LW $R7 $R19 0
0b00000000011100100001110000000100000, L114. ADD $R7 $R4 $R7
0b10001100111110100000000000000000000, L115. LW $R7 $R20 0 \\ End of Algorithm
    4.15
0b00010101001000000000000010001001, 1116. BNE $R9 $R0 137 \\ Start of
    Algorithm 4.16
0b00000000000000000000000000000000000, L117. SLL $R0 $R0 $R0
0b00000010010101001010100000110000, L118. ADDS $R18 $R20 $R21
0b00000010011101011001100000100010, L119. SUB $R19 $R21 $R19
0b00000010011001110101000000110010, L120. ADDSS $R19 $R19 $R21
0b00000010101100101001000000100000, L121. ADD $R21 $R18 $R18
0b00000010011000010001000000100000, L122. ADD $R19 $R0 $R17
0b10101101000100100000000000000000, L123. SW $R8 $R18 0
0b00000001000011000101100000100010, L124. SUB $R8 $R12 $R11 \\ Start of
    Algorithm 4.17
0b00000001011001100101100000111000, L125. DWTA $R11 $R6 $R11
0b00000001011101001011000000100000, L126. ADD $R11 $R26 $R11 \\ End of
    Algorithm 4.17
0b10101101010110010000000000000000, L127. SW $R11 $R18 0
0b00000001000001000100000000100000, L128. ADD $R8 $R4 $R8
0b10101101000100110000000000000000, L129. SW $R8 $R19 0
0b00000001000011000101100000100010, L130. SUB $R8 $R12 $R11 \\ Start of
    Algorithm 4.17
0b00000001011001100101100000111000, L131. DWTA $R11 $R6 $R11
0b00000001011101001011000000100000, L132. ADD $R11 $R26 $R11 \\ End of
    Algorithm 4.17
0b10101101011100100000000000000000, L133. SW $R11 $R19 0
0b00000001000001000100000000100000, L134. ADD $R8 $R4 $R8
0b00010000000000000000000010101101, L135. BEQ $R0 $R0 173
0b00000000000000000000000000000000, L136. SLL $R0 $R0 $R0
0b00010101001000000000000010011100, L137. BNE $R9 $R16 156
0b00000000000000000000000000000000, L138. SLL $R0 $R0 $R0
0b00000010010100101010100000110000, L139. ADDS $R18 $R18 $R21
0b00000010011101011001100000100010, L140. SUB $R19 $R21 $R19
0b00000010011001110101000000110010, L141. ADDSS $R17 $R19 $R21
0b00000010101100101001000000100000, L142. ADD $R21 $R18 $R18
0b10101101000100100000000000000000, L143. SW $R8 $R18 0
0b00000001000011000101100000100010, L144. SUB $R8 $R12 $R11 \\ Start of
    Algorithm 4.17
0b00000001011001100101100000111000, L145. DWTA $R11 $R6 $R11
0b00000001011101001011000000100000, L146. ADD $R11 $R26 $R11 \\ End of
    Algorithm 4.17
0b10101101011100100000000000000000, L147. SW $R11 $R18 0
0b00000001000001000100000000100000, L148. ADD $R8 $R4 $R8
0b10101101000100110000000000000000, L149. SW $R8 $R19 0
0b00000001000011000101100000100010, L150. SUB $R8 $R12 $R11 \\ Start of
    Algorithm 4.17
0b00000001011001100101100000111000, L151. DWTA $R11 $R6 $R11
0b00000001011101001011000000100000, L152. ADD $R11 $R26 $R11 \\ End of
    Algorithm 4.17
0b10101101011100110000000000000000, L153. SW $R11 $R19 0
0b00010000000000000000000010101101, L154. BEQ $R0 $R0 173
0b00000000000000000000000000000000, L155. SLL $R0 $R0 $R0
0b00000010010101001010100000110000, L156. ADDS $R18 $R20 $R21
0b00000010011101011001100000100010, L157. SUB $R19 $R21 $R19
0b00000010011001110101000000110010, L158. ADDSS $R17 $R19 $R21
0b00000010101100101001000000100000, L159. ADD $R21 $R18 $R18
0b00000010011000010001000000100000, L160. ADD $R19 $R0 $R17
0b10101101000100100000000000000000, L161. SW $R8 $R18 0
0b00000001000011000101100000100010, L162. SUB $R8 $R12 $R11 \\ Start of
    Algorithm 4.17
0b00000001011001100101100000111000, L163. DWTA $R11 $R6 $R11
```

```
0b00000000101111010010110000100000, L164. ADD $R11 $R26 $R11 \\ End of
     Algorithm 4.17
0b10101101011100100000000000000000, L165. SW $R11 $R18 0
0b00000000100000100010000000100000, L166. ADD $R8 $R4 $R8
0b10101101000100110000000000000000, L167. SW $R8 $R19 0
0b00000001000011000101100000100010, L168. SUB $R8 $R12 $R11 \\ Start of
     Algorithm 4.17
0b00000000101100110010110000111000, L169. DWTA $R11 $R6 $R11
0b00000000101111010010110000100000, L170. ADD $R11 $R26 $R11 \\ End of
     Algorithm 4.17
0b10101101011100110000000000000000, L171. SW $R11 $R19 0
0b00000000100000100010000000100000, L172. ADD $R8 $R4 $R8 \\ End of
     Algorithm 4.16
0b00000001001000100100100000100000, L173. ADD $R9 $R2 $R9
0b00010000000000000000000001100101, L174. BEQ $R0 $R0 101 \\ end while for
     $R9 < $R3 do
0b00000000000000000000000000000000, L175. SLL $R0 $R0 $R0
0b00000000000000000100100000100000, L176. ADD $R0 $R0 $R9
0b00000001010000010101000000100000, L177. ADD $R10 $R1 $R10
0b00000001010000000011100000100000, L178. ADD $R10 $R0 $R7
0b00000000111011000011100000100000, L179. ADD $R7 $R12 $R7
0b00000001010000000100000000100000, L180. ADD $R10 $R0 $R8
0b00000001000011000100000000100000, L181. ADD $R8 $R12 $R8
0b00010000000000000000000001100010, L182. BEQ $R0 $R0 98 \\ end while for
     $R10 < $R4 do
0b00000000000000000000000000000000, L183. SLL $R0 $R0 $R0
0b00000000011000000001100000000010, L184. SRL $R3 $R0 $R3 \\ Start of
     Algorithm 4.18
0b00000000010000000010000000000010, L185. SRL $R4 $R0 $R4
0b00000000010000000010100000000010, L186. SRL $R5 $R0 $R5
0b00000000010000000010100000000010, L187. SRL $R5 $R0 $R5
0b00000000110000010001100000100000, L188. ADD $R6 $R1 $R6
0b00100000000101100000000000000101, L189. ADDI $R0 $R22 5
0b00000000110101101011000000101010, L190. SLT $R6 $R22 $R22
0b00010110110000010000000011001101, L191. BNE $R22 $R1 205
0b00000000000000000000000000000000, L192. SLL $R0 $R0 $R0
0b00100000000110000010000000000000, L193. ADDI $R0 $R24 2048
0b00100000000110010001000000000000, L194. ADDI $R0 $R25 4096
0b00000000000000001101100000100000, L195. ADD $R0 $R0 $R27
0b10001111000101110000000000000000, L196. LW $R24 $R22 0
0b10101111100110110000000000000000, L197. SW $R25 $R22 0
0b00000011000000011100000000100000, L198. ADD $R24 $R1 $R24
0b00000011001000011100100000100000, L199. ADD $R25 $R1 $R25
0b00000011011000011101100000100000, L200. ADD $R27 $R1 $R27
0b00010111011001010000000011000100, L201. BNE $R27 $R5 196
0b00000000000000000000000000000000, L202. SLL $R0 $R0 $R0
0b00010000000000000000000000010010, L203. BEQ $R0 $R0 18
0b00000000000000000000000000000000, L204. SLL $R0 $R0 $R0
0b00000000000000000000000000000000, L205. NOP                      \\ End of
     Algorithm 4.18
     }
     with {block = 1};

     //*********************************************************//
     // Signal Lines Declarations (requires no actual memory storage)
     //*********************************************************//
     // Instruction Fetch (IF) Stage ------------------------------
     signal unsigned int 16           Signal_IF_PC;              //
         Intermediate signal: PC value
     signal unsigned int 16           Signal_IF_PC_Mux;          //
         Intermediate signal: Output from PC_Mux
     signal unsigned int 16           Signal_IF_PC_Increment;    //
         Intermediate signal: PC + 1
     signal unsigned int 32           Signal_IF_Inst;            //
         Intermediate signal: Instructions from INST_MEMORY
```

```
    // Instruction Decode (ID) Stage -----------------------------
    signal unsigned int 32        Signal_ID_Inst;                  //
        Intermediate signal: Instruction read from IFID Pipeline
    signal unsigned int 16        Signal_ID_ReadReg1;             //
        Intermediate signal: Read Data 1 out from Register (before
        forwarding)
    signal unsigned int 16        Signal_ID_ReadReg2;             //
        Intermediate signal: Read Data 2 out from Register(before
        forwarding)
    signal unsigned int 16        Signal_ID_ReadData1;            //
        Intermediate signal: Read Data 1 out from Register (after
        forwarding)
    signal unsigned int 16        Signal_ID_ReadData2;            //
        Intermediate signal: Read Data 2 out from Register (after
        forwarding)
    //signal unsigned int 32         Signal_ID_SignExt;            //
        Intermediate signal: Extended signal from 16 to 32 bits

    // Branch Unit
    signal unsigned int 1         Signal_ID_Comp;                 //
        Intermediate signal: Comparator output 1 - equal, 0 - not equal
    signal unsigned int 1         Signal_ID_XOROut;               //
        Intermediate signal: Output from Branch Unit into ID AND gate

    // Operation Control Signals
    signal unsigned int 1         Signal_ID_RegDst;            // Control
        signal: Controls Register Destination
    signal unsigned int 1         Signal_ID_ALUSrc;            // Control
        signal: Controls Source into ALU
    signal unsigned int 1         Signal_ID_ALUOp1;            // Control
        signal: Controls ALU Operation (bit 1)
    signal unsigned int 1         Signal_ID_ALUOp0;            // Control
        signal: Controls ALU Operation (bit 0)
    signal unsigned int 1         Signal_ID_Branch;            // Control
        signal: Generates branch signal on branch instructions
    signal unsigned int 1         Signal_ID_MemRead;           // Control
        signal: Allows memory read when asserted
    signal unsigned int 1         Signal_ID_MemWrite;          // Control
        signal: Allows memory write when asserted
    signal unsigned int 1         Signal_ID_MemtoReg;          // Control
        signal: Allows memory read data to Register when asserted
    signal unsigned int 1         Signal_ID_RegWrite;          // Control
        signal: Allows Register write when asserted

    // ID Mux A and B Control Signals
    signal unsigned int 1         Signal_ID_MuxA_Ctrl;         // Control
        signal: Forwarding from WB stage for ReadData1
    signal unsigned int 1         Signal_ID_MuxB_Ctrl;         // Control
        signal: Forwarding from WB stage for ReadData2

    // IF Stage Control Signals from ID stage
    signal unsigned int 1         Signal_ID_PCMux_Ctrl;        // Control
        signal: IF PCMUX control
    signal unsigned int 1         Signal_ID_PCWrite;           // Control
        signal: IF PCWrite control
    signal unsigned int 1         Signal_ID_IFIDWrite;         // Control
        signal: Pipeline Register write control

    // Hazard Detection Unit
    signal unsigned int 1         Signal_ID_HazardMux_Ctrl;   // Control
        signal: Selects zero for all controls and for Rs and Rt into the ID
        /EXMEM pipeline

    // Hazard Selection Outputs
    signal unsigned int 1         Signal_ID_Hzd_RegDst;        // Control
        signal: Controls Register Destination
```

```
    signal unsigned int 1              Signal_ID_Hzd_ALUSrc;        // Control
        signal: Controls Source into ALU
    signal unsigned int 1              Signal_ID_Hzd_ALUOp1;        // Control
        signal: Controls ALU Operation (bit 1)
    signal unsigned int 1              Signal_ID_Hzd_ALUOp0;        // Control
        signal: Controls ALU Operation (bit 0)
    signal unsigned int 1              Signal_ID_Hzd_MemRead;       // Control
        signal: Allows memory read when asserted
    signal unsigned int 1              Signal_ID_Hzd_MemWrite;      // Control
        signal: Allows memory write when asserted
    signal unsigned int 1              Signal_ID_Hzd_MemtoReg;      // Control
        signal: Allows memory read data to Register when asserted
    signal unsigned int 1              Signal_ID_Hzd_RegWrite;      // Control
        signal: Allows Register write when asserted

    signal unsigned int 5              Signal_ID_Hzd_Rs;            //
        Intermediate signal: value from Rs field(R-format)/also for
        forwarding
    signal unsigned int 5              Signal_ID_Hzd_Rt;            //
        Intermediate signal: value from Rt field(R-format)/Rd field(I-
        format)
    signal unsigned int 5              Signal_ID_Hzd_Rd;            //
        Intermediate signal: value from Rd field(R-format)

// Instruction Execution and Memory (EXMEM) Stage ------------
    signal unsigned int 16             Signal_EXMEM_ReadData1;      //
        Intermediate signal: Read Data 1 out from Register
    signal unsigned int 16             Signal_EXMEM_ReadData2;      //
        Intermediate signal: Read Data 2 out from Register
    signal unsigned int 16             Signal_EXMEM_ImmVal;         //
        Intermediate signal: Immediate value from Inst[15:0]
    signal unsigned int 16             Signal_EXMEM_ALUIn2;         //
        Intermediate signal: ALU input 2
    signal unsigned int 16             Signal_EXMEM_ALU_Result;     //
        Intermediate signal: ALU result
    signal unsigned int 16             Signal_EXMEM_MemData;        //
        Intermediate signal: Data from DATAMEM
    signal unsigned int 16             Signal_EXMEM_WBData;         //
        Intermediate signal: Writeback Data

    signal unsigned int 5              Signal_EXMEM_Rs;             //
        Intermediate signal: value from Rs field(R-format)/also for
        forwarding
    signal unsigned int 5              Signal_EXMEM_Rt;             //
        Intermediate signal: value from Rt field(R-format)/Rd field(I-
        format)
    signal unsigned int 5              Signal_EXMEM_Rd;             //
        Intermediate signal: value from Rd field(R-format)
    signal unsigned int 5              Signal_EXMEM_Regd;           //
        Intermediate signal: selected destination register

    signal unsigned int 1              Signal_EXMEM_RegDst;         // Control
        signal: Controls Register Destination
    signal unsigned int 1              Signal_EXMEM_ALUSrc;         // Control
        signal: Controls Source into ALU
    signal unsigned int 1              Signal_EXMEM_ALUOp1;         // Control
        signal: Controls ALU Operation (bit 1)
    signal unsigned int 1              Signal_EXMEM_ALUOp0;         // Control
        signal: Controls ALU Operation (bit 0)
    signal unsigned int 1              Signal_EXMEM_MemRead;        // Control
        signal: Allows memory read when asserted
    signal unsigned int 1              Signal_EXMEM_MemWrite;       // Control
        signal: Allows memory write when asserted
    signal unsigned int 1              Signal_EXMEM_MemtoReg;       // Control
        signal: Allows memory read data to Register when asserted
    signal unsigned int 1              Signal_EXMEM_RegWrite;       // Control
        signal: Allows Register write when asserted
```

```
signal unsigned int 16          Signal_EXMEM_FA_Mux;         //
    Intermediate signal: output from Forwarding MUX A, also ALUIn1
signal unsigned int 16          Signal_EXMEM_FB_Mux;         //
    Intermediate signal: output from Forwarding MUX B

signal unsigned int 1           Signal_EXMEM_FA;             // Control
    signal: Forwarding MUX A data selection
signal unsigned int 1           Signal_EXMEM_FB;             // Control
    signal: Forwarding MUX B data selection

signal unsigned int 4           Signal_EXMEM_ALU_Ctrl;       // Control
    signal: Determines specific ALU operation
signal unsigned int 1           Signal_EXMEM_ALU_Ctrl_B3;    // Control
    signal: Determines specific ALU operation (bit-3)
signal unsigned int 1           Signal_EXMEM_ALU_Ctrl_B2;    // Control
    signal: Determines specific ALU operation (bit-2)
signal unsigned int 1           Signal_EXMEM_ALU_Ctrl_B1;    // Control
    signal: Determines specific ALU operation (bit-1)
signal unsigned int 1           Signal_EXMEM_ALU_Ctrl_B0;    // Control
    signal: Determines specific ALU operation (bit-0)


// Write Back (WB) Stage
signal unsigned int 1           Signal_WB_MemtoReg;          // Control
    signal: Allows memory read data to Register when asserted
signal unsigned int 1           Signal_WB_RegWrite;          // Control
    signal: Allows Register write when asserted
signal unsigned int 5           Signal_WB_Rd;                //
    Intermediate signal: Register destination addr
signal unsigned int 16          Signal_WB_ALUData;           //
    Intermediate signal: ALU Result
signal unsigned int 16          Signal_WB_MemData;           //
    Intermediate signal: Memory Read Data
signal unsigned int 16          Signal_WB_Mem_Data;          //
    Intermediate signal: Memory out
signal unsigned int 16          Signal_WB_ALU_Result;        //
    Intermediate signal: ALU result
signal unsigned int 16          Signal_WB_Data;              //
    Intermediate signal: Write back data


//***********************************************************//
// Components Declarations (requires actual memory storage)
//***********************************************************//
// 32-bit General Purpose Registers
unsigned int 16         Register[32];                        // 32
    general purpose registers

// Misc Registers
unsigned int 1          RUN;                                 // 1 - Run
    processor / 0 - Terminate
unsigned int 32         ClockCycle;                          // Clock
    cycle counter

// IF/ID Stage Registers
unsigned int 16         PC;                                  // Program
    Counter
unsigned int 32         IFID_Inst;                           // IF/ID
    Register: Holds instructions from INST_MEMORY

// ID/EXMEM Stage Registers
unsigned int 1          IDEXMEM_WB_MemtoReg;                 // ID/EXMEM
    Register: Holds WB MemtoReg control signal
unsigned int 1          IDEXMEM_WB_RegWrite;                 // ID/EXMEM
    Register: Holds WB RegWrite control signal
unsigned int 1          IDEXMEM_MEM_MemRead;                 // ID/EXMEM
    Register: Holds MEM MemRead control signal
unsigned int 1          IDEXMEM_MEM_MemWrite;                // ID/EXMEM
    Register: Holds MEM MemWrite control signal
```

```
unsigned int 1            IDEXMEM_EX_RegDst;                    // ID/EXMEM
     Register: Holds EX RegDst control signal
unsigned int 1            IDEXMEM_EX_ALUOp1;                    // ID/EXMEM
     Register: Holds EX ALUOp1 control signal
unsigned int 1            IDEXMEM_EX_ALUOp0;                    // ID/EXMEM
     Register: Holds EX ALUOp0 control signal
unsigned int 1            IDEXMEM_EX_ALUSrc;                    // ID/EXMEM
     Register: Holds EX ALUSrc control signal

unsigned int 16           IDEXMEM_ReadData1;                    // ID/EXMEM
     Register: Holds read data 1 from Register
unsigned int 16           IDEXMEM_ReadData2;                    // ID/EXMEM
     Register: Holds read data 2 from Register
unsigned int 16           IDEXMEM_ImmVal;                       // ID/EXMEM
     Register: Holds immediate value for Instuction[15:0]
unsigned int 5            IDEXMEM_Rs;                           // ID/EXMEM
     Register: Holds value from Rs field(R-format)/also for forwarding
unsigned int 5            IDEXMEM_Rt;                           // ID/EXMEM
     Register: Holds value from Rt field(R-format)/Rd field(I-format)
unsigned int 5            IDEXMEM_Rd;                           // ID/EXMEM
     Register: Holds value from Rd field(R-format)

// EXMEM/WB Stage Registers
unsigned int 1            EXMEMWB_WB_RegWrite;                  // EXMEM/WB
     Register: Holds WB RegWrite control signal
unsigned int 1            EXMEMWB_WB_MemtoReg;                  // EXMEM/WB
     Register: Holds WB MemtoReg control signal
unsigned int 16           EXMEMWB_WB_ALUData;                   // EXMEM/WB
     Register: Holds ALU Data
unsigned int 16           EXMEMWB_WB_MemData;                   // EXMEM/WB
     Register: Holds Memory Read Data
unsigned int 5            EXMEMWB_Rd;                           // EXMEM/WB
     Register: Holds register destination addr

// Components Initialization
RUN            = 1;          // Set RUN flag to 1
ClockCycle     = 0;          // ClockCycle counter starts at 0
PC             = 0;          // PC starts at 0
Register[0]    = 0;
Register[1]    = 1;
Register[2]    = 2;

// Processor Core
while (RUN == 1) // run processor
{
    par // begin parallel execution
    {
        //----------------------------------------------------------------
        // Intermediate signal values
        //----------------------------------------------------------------

        // Stage 1: Instruction Fetch ------------------------------------
        Signal_IF_PC           = PC;
        Signal_IF_PC_Increment = Signal_IF_PC + 1;

        // PC source selection
        if(Signal_ID_PCMux_Ctrl == 0) // if no branch occurs
        {
            Signal_IF_PC_Mux   = Signal_IF_PC_Increment;
        }
        else // if branch occurs
        {
            Signal_IF_PC_Mux   = Signal_ID_Inst[15:0];
        }

        // Instruction fetching from INST_MEMORY
        Signal_IF_Inst         = INST_MEMORY[Signal_IF_PC[10:0]];
```

```
// Stage 2: Instruction Decode --------------------------------
Signal_ID_Inst          = IFID_Inst;
Signal_ID_ReadReg1      = Register[Signal_ID_Inst[25:21]];
Signal_ID_ReadReg2      = Register[Signal_ID_Inst[20:16]];

Signal_ID_RegDst        = (~Signal_ID_Inst[29]) & (~
    Signal_ID_Inst[26]);
Signal_ID_ALUSrc        = Signal_ID_Inst[29] | ((~Signal_ID_Inst
    [28]) & Signal_ID_Inst[26]);
Signal_ID_MemtoReg      = Signal_ID_Inst[26];
Signal_ID_RegWrite      = ((~Signal_ID_Inst[29]) & (~
    Signal_ID_Inst[28])) |
                            ((~Signal_ID_Inst[28]) & (~
                                Signal_ID_Inst[26]));
Signal_ID_MemRead       = (~Signal_ID_Inst[29]) & (~
    Signal_ID_Inst[28]) & Signal_ID_Inst[26];
Signal_ID_MemWrite      = Signal_ID_Inst[29] & Signal_ID_Inst
    [26];
Signal_ID_ALUOp1        = (~Signal_ID_Inst[29]) & (~
    Signal_ID_Inst[28]) & (~Signal_ID_Inst[26]);
Signal_ID_ALUOp0        = Signal_ID_Inst[28];
Signal_ID_Branch        = Signal_ID_Inst[28];

// Branch Unit consist of equality comparator and a XOR gate
// comparator logic for reducing branch hazard
// still requires one NOP instructions after branch instruction
    (conventional is 3 NOPs)
if(Signal_ID_ReadData1[15:0] == Signal_ID_ReadData2[15:0]) // if
    both data values are equal
{Signal_ID_Comp = 1;}
else{Signal_ID_Comp = 0;}

// Generate Branch Unit output
Signal_ID_XOROut        = Signal_ID_Comp ^ Signal_ID_Inst[26];

 // Generate MUX signal for selection of new PC value
Signal_ID_PCMux_Ctrl    = Signal_ID_Branch & Signal_ID_XOROut;

// Forwarding logic
if((Signal_WB_RegWrite == 1) && (Signal_WB_Rd != 0) && (
    Signal_WB_Rd == Signal_ID_Inst[25:21]))
{Signal_ID_MuxA_Ctrl = 1;}
else{Signal_ID_MuxA_Ctrl = 0;}

if((Signal_WB_RegWrite == 1) && (Signal_WB_Rd != 0) && (
    Signal_WB_Rd == Signal_ID_Inst[20:16]))
{Signal_ID_MuxB_Ctrl = 1;}
else{Signal_ID_MuxB_Ctrl = 0;}

// Forwarding ID Mux A data selection
if(Signal_ID_MuxA_Ctrl == 1)
{Signal_ID_ReadData1 = Signal_WB_Data;}
else{Signal_ID_ReadData1 = Signal_ID_ReadReg1;}

// Forwarding ID Mux B data selection
if(Signal_ID_MuxB_Ctrl == 1)
{Signal_ID_ReadData2 = Signal_WB_Data;}
else{Signal_ID_ReadData2 = Signal_ID_ReadReg2;}

// Register write
if(Signal_WB_RegWrite == 1)
{Register[Signal_WB_Rd] = Signal_WB_Data;}
else
{delay;}
```

```
            // Hazard Detection Unit
            if((Signal_ID_Branch == 1) && (((Signal_ID_Inst[25:21] ==
                Signal_EXMEM_Rs) && (Signal_ID_Inst[25:21] != 0) && (
                Signal_ID_Inst[25:21] != 1))
                || ((Signal_ID_Inst[25:21] == Signal_EXMEM_Regd) && (
                    Signal_ID_Inst[25:21] != 0) && (Signal_ID_Inst[25:21]
                    != 1))
                || ((Signal_ID_Inst[20:16] == Signal_EXMEM_Rs) && (
                    Signal_ID_Inst[20:16] != 0) && (Signal_ID_Inst[20:16]
                    != 1))
                || ((Signal_ID_Inst[20:16] == Signal_EXMEM_Regd) && (
                    Signal_ID_Inst[20:16] != 0) && (Signal_ID_Inst[20:16]
                    != 1))))
            {
                par
                {
                    Signal_ID_PCWrite           = 0;
                    Signal_ID_IFIDWrite         = 0;
                    Signal_ID_HazardMux_Ctrl = 1;
                }
            }
            else // no hazard
            {
                par
                {
                    Signal_ID_PCWrite           = 1;
                    Signal_ID_IFIDWrite         = 1;
                    Signal_ID_HazardMux_Ctrl = 0;
                }
            }

            if(Signal_ID_HazardMux_Ctrl == 0)
            {
                par
                {
                    Signal_ID_Hzd_Rs            = Signal_ID_Inst[25:21];
                    Signal_ID_Hzd_Rt            = Signal_ID_Inst[20:16];
                    Signal_ID_Hzd_Rd            = Signal_ID_Inst[15:11];
                }
            }
            else
            {
                par
                {
                    Signal_ID_Hzd_Rs            = 0;
                    Signal_ID_Hzd_Rt            = 0;
                    Signal_ID_Hzd_Rd            = 0;
                }
            }

            if(Signal_ID_HazardMux_Ctrl == 0)
            {
                par
                {
                    Signal_ID_Hzd_MemtoReg      = Signal_ID_MemtoReg;
                    Signal_ID_Hzd_RegWrite      = Signal_ID_RegWrite;
                    Signal_ID_Hzd_MemRead       = Signal_ID_MemRead;
                    Signal_ID_Hzd_MemWrite      = Signal_ID_MemWrite;
                    Signal_ID_Hzd_RegDst        = Signal_ID_RegDst;
                    Signal_ID_Hzd_ALUOp1        = Signal_ID_ALUOp1;
                    Signal_ID_Hzd_ALUOp0        = Signal_ID_ALUOp0;
                    Signal_ID_Hzd_ALUSrc        = Signal_ID_ALUSrc;
                }
            }
            else
            {
                par
```

```
        {
           Signal_ID_Hzd_MemtoReg   = 0;
           Signal_ID_Hzd_RegWrite   = 0;
           Signal_ID_Hzd_MemRead    = 0;
           Signal_ID_Hzd_MemWrite   = 0;
           Signal_ID_Hzd_RegDst     = 0;
           Signal_ID_Hzd_ALUOp1     = 0;
           Signal_ID_Hzd_ALUOp0     = 0;
           Signal_ID_Hzd_ALUSrc     = 0;
        }
      }

      // Stage 3: Execution -------------------------------------
      Signal_EXMEM_ReadData1     = IDEXMEM_ReadData1;
      Signal_EXMEM_ReadData2     = IDEXMEM_ReadData2;
      Signal_EXMEM_ImmVal        = IDEXMEM_ImmVal;

      Signal_EXMEM_Rs            = IDEXMEM_Rs;
      Signal_EXMEM_Rt            = IDEXMEM_Rt;
      Signal_EXMEM_Rd            = IDEXMEM_Rd;

      Signal_EXMEM_MemtoReg      = IDEXMEM_WB_MemtoReg;
      Signal_EXMEM_RegWrite      = IDEXMEM_WB_RegWrite;
      Signal_EXMEM_MemRead       = IDEXMEM_MEM_MemRead;
      Signal_EXMEM_MemWrite      = IDEXMEM_MEM_MemWrite;
      Signal_EXMEM_RegDst        = IDEXMEM_EX_RegDst;
      Signal_EXMEM_ALUOp1        = IDEXMEM_EX_ALUOp1;
      Signal_EXMEM_ALUOp0        = IDEXMEM_EX_ALUOp0;
      Signal_EXMEM_ALUSrc        = IDEXMEM_EX_ALUSrc;

      // Data forwarding logic
      if((Signal_WB_RegWrite == 1) && (Signal_WB_Rd != 0) && (
          Signal_WB_Rd == Signal_EXMEM_Rs))
      {Signal_EXMEM_FA = 1;}
      else
      {Signal_EXMEM_FA = 0;}

      if((Signal_WB_RegWrite == 1) && (Signal_WB_Rd != 0) && (
          Signal_WB_Rd == Signal_EXMEM_Rt))
      {Signal_EXMEM_FB = 1;}
      else
      {Signal_EXMEM_FB = 0;}

      // Forwarding A MUX data selection
      if(Signal_EXMEM_FA == 0)          // $R1 data is available (no
          dependency)
      {Signal_EXMEM_FA_Mux = Signal_EXMEM_ReadData1;}
      else
      {Signal_EXMEM_FA_Mux = Signal_WB_Data;}
      {delay;}

      // Forwarding B MUX data selection
      if(Signal_EXMEM_FB == 0)          // $R2 data is available (no
          dependency)
      {Signal_EXMEM_FB_Mux = Signal_EXMEM_ReadData2;}
      else
      {Signal_EXMEM_FB_Mux = Signal_WB_Data;}
      {delay;}

      // ALU Input 2 data selection
      if(Signal_EXMEM_ALUSrc == 0)                    // R-FORMAT, input
          from rt field
      {Signal_EXMEM_ALUIn2 = Signal_EXMEM_FB_Mux;}
      else                                            // I-FORMAT, input
          from const/addr field
      {Signal_EXMEM_ALUIn2 = Signal_EXMEM_ImmVal;}
```

```
// ALU Control
Signal_EXMEM_ALU_Ctrl_B3   = (Signal_EXMEM_ALUOp1 &
    Signal_EXMEM_ImmVal[4]) | (Signal_EXMEM_ALUOp1 &
                                (~Signal_EXMEM_ImmVal[5]) &
                                    Signal_EXMEM_ImmVal[2]);
Signal_EXMEM_ALU_Ctrl_B2   = (Signal_EXMEM_ALUOp1 &
    Signal_EXMEM_ImmVal[3]) | (Signal_EXMEM_ALUOp1 &
                                (~Signal_EXMEM_ImmVal[5]) & (~
                                    Signal_EXMEM_ImmVal[2])) |
                                (Signal_EXMEM_ALUOp1 &
                                Signal_EXMEM_ImmVal[2] &
                                    Signal_EXMEM_ImmVal[1]);
Signal_EXMEM_ALU_Ctrl_B1   = (Signal_EXMEM_ALUOp1 &
    Signal_EXMEM_ImmVal[4] & (~Signal_EXMEM_ImmVal[3])) |
                                (Signal_EXMEM_ALUOp1 & (~
                                Signal_EXMEM_ImmVal[5]) &
                                Signal_EXMEM_ImmVal[1]) |
                                (Signal_EXMEM_ALUOp1 &
                                Signal_EXMEM_ImmVal[5] &
                                Signal_EXMEM_ImmVal[2] &
                                (~Signal_EXMEM_ImmVal[1])) | (
                                Signal_EXMEM_ALUOp1 &
                                Signal_EXMEM_ImmVal[3] &
                                Signal_EXMEM_ImmVal[1]) | (
                                Signal_EXMEM_ALUOp1 & (~
                                Signal_EXMEM_ImmVal[3]) &
                                (~Signal_EXMEM_ImmVal[2]) &
                                Signal_EXMEM_ImmVal[0]);
Signal_EXMEM_ALU_Ctrl_B0   = (Signal_EXMEM_ALUOp1 &
    Signal_EXMEM_ImmVal[0]) | (Signal_EXMEM_ALUOp1 &
                                (~Signal_EXMEM_ImmVal[5]) & (~
                                Signal_EXMEM_ImmVal[2]) &
                                (~Signal_EXMEM_ImmVal[1])) | (
                                Signal_EXMEM_ALUOp1 &
                                Signal_EXMEM_ImmVal[5] &
                                (~Signal_EXMEM_ImmVal[4]) & (~
                                Signal_EXMEM_ImmVal[2]) &
                                Signal_EXMEM_ImmVal[1]) | (
                                Signal_EXMEM_ALUOp1 &
                                Signal_EXMEM_ImmVal[5] &
                                (~Signal_EXMEM_ImmVal[3]) & (~
                                Signal_EXMEM_ImmVal[2]) &
                                Signal_EXMEM_ImmVal[1]);
Signal_EXMEM_ALU_Ctrl     = Signal_EXMEM_ALU_Ctrl_B3 @
    Signal_EXMEM_ALU_Ctrl_B2
                                @ Signal_EXMEM_ALU_Ctrl_B1 @
                                Signal_EXMEM_ALU_Ctrl_B0;

// ALU
if(Signal_EXMEM_ALU_Ctrl == 0b0000)        // ADD
{
    Signal_EXMEM_ALU_Result = 0@((unsigned)(((signed)
        Signal_EXMEM_FA_Mux[15:0]) + ((signed)
        Signal_EXMEM_ALUIn2[15:0])));
}
else if(Signal_EXMEM_ALU_Ctrl == 0b0001)   // SUB
{
    Signal_EXMEM_ALU_Result = 0@((unsigned)(((signed)
        Signal_EXMEM_FA_Mux[15:0]) - ((signed)
        Signal_EXMEM_ALUIn2[15:0])));
}
else if(Signal_EXMEM_ALU_Ctrl == 0b0010)   // AND
{
    Signal_EXMEM_ALU_Result = 0@((unsigned)(((signed)
        Signal_EXMEM_FA_Mux[15:0]) & ((signed)
        Signal_EXMEM_ALUIn2[15:0])));
}
```

```
else if(Signal_EXMEM_ALU_Ctrl == 0b0011)  // OR
{
    Signal_EXMEM_ALU_Result = 0@((unsigned)(((signed)
        Signal_EXMEM_FA_Mux[15:0]) | ((signed)
        Signal_EXMEM_ALUIn2[15:0])));
}
else if(Signal_EXMEM_ALU_Ctrl == 0b0100)  // NOT
{
    Signal_EXMEM_ALU_Result = 0@((unsigned)(~((signed)
        Signal_EXMEM_FA_Mux[15:0])));
}
else if(Signal_EXMEM_ALU_Ctrl == 0b0101)  // SLL
{
    Signal_EXMEM_ALU_Result = 0@((unsigned)(((signed)
        Signal_EXMEM_FA_Mux[15:0]) << 1));
}
else if(Signal_EXMEM_ALU_Ctrl == 0b0110)  // SRL
{
    Signal_EXMEM_ALU_Result = 0@((unsigned)(((signed)
        Signal_EXMEM_FA_Mux[15:0]) >> 1));
}
else if(Signal_EXMEM_ALU_Ctrl == 0b0111)  // SLT
{
    if(((signed)Signal_EXMEM_FA_Mux[15:0]) < ((signed)
        Signal_EXMEM_ALUIn2[15:0]))
    {Signal_EXMEM_ALU_Result = 1;}
    else
    {Signal_EXMEM_ALU_Result = 0;}
}
else if(Signal_EXMEM_ALU_Ctrl == 0b1000)  // ABS
{
    // To perform absolute, invert all bits is bit 12 is 1, and
        then add 1 to the result
    if(Signal_EXMEM_FA_Mux[15] == 1)
    {
        Signal_EXMEM_ALU_Result = 0@((~Signal_EXMEM_FA_Mux
            [15:0]) + 1);
    }
    else
    {
        Signal_EXMEM_ALU_Result = 0@(Signal_EXMEM_FA_Mux[15:0]);
    }
}
else if(Signal_EXMEM_ALU_Ctrl == 0b1001)  // ABSSUB
{
    // if second field is larger than the first field there will
        be a negative result before absolute
    if(((signed)Signal_EXMEM_FA_Mux[15:0]) < ((signed)
        Signal_EXMEM_ALUIn2[15:0]))
    {
        Signal_EXMEM_ALU_Result = 0@((~((unsigned)(((signed)
            Signal_EXMEM_FA_Mux[15:0]) - ((signed)
            Signal_EXMEM_ALUIn2[15:0])))) + 1);
    }
    else // normal subtraction
    {
        Signal_EXMEM_ALU_Result = 0@((unsigned)(((signed)
            Signal_EXMEM_FA_Mux[15:0]) - ((signed)
            Signal_EXMEM_ALUIn2[15:0])));
    }
}
else if(Signal_EXMEM_ALU_Ctrl == 0b1010)  // ADDS
{
    Signal_EXMEM_ALU_Result = 0@((unsigned)(((signed)
        Signal_EXMEM_FA_Mux[15:0] + (signed)Signal_EXMEM_ALUIn2
        [15:0]) >> 1));
}
```

```
        else if(Signal_EXMEM_ALU_Ctrl == 0b1011)  // ADDSS
        {
            Signal_EXMEM_ALU_Result = 0@((unsigned)(((signed)
                  Signal_EXMEM_FA_Mux[15:0] + (signed)Signal_EXMEM_ALUIn2
                  [15:0] + (signed)2) >> 2));
        }
        else if(Signal_EXMEM_ALU_Ctrl == 0b1100)  // DWT ADDRESS
              REARRANGE
        {
            if(Signal_EXMEM_ALUIn2[14:0] == 1) // Lv1
            {
                Signal_EXMEM_ALU_Result = 0[15:11] @ Signal_EXMEM_FA_Mux
                      [7] @ Signal_EXMEM_FA_Mux[0] @
                                        Signal_EXMEM_FA_Mux[10:8] @
                                        Signal_EXMEM_FA_Mux
                                        [6:1];
            }
            else if(Signal_EXMEM_ALUIn2[14:0] == 2) // Lv3
            {
                Signal_EXMEM_ALU_Result = 0[15:9] @ Signal_EXMEM_FA_Mux
                      [6] @ Signal_EXMEM_FA_Mux[0] @
                                        Signal_EXMEM_FA_Mux[8:7] @
                                        Signal_EXMEM_FA_Mux
                                        [5:1];
            }
            else if(Signal_EXMEM_ALUIn2[14:0] == 3) // Lv3
            {
                Signal_EXMEM_ALU_Result = 0[15:7] @ Signal_EXMEM_FA_Mux
                      [5] @ Signal_EXMEM_FA_Mux[0] @
                                        Signal_EXMEM_FA_Mux[6] @
                                        Signal_EXMEM_FA_Mux
                                        [4:1];
            }
            else if(Signal_EXMEM_ALUIn2[14:0] == 4) // Lv4
            {
                Signal_EXMEM_ALU_Result = 0[15:5] @ Signal_EXMEM_FA_Mux
                      [4] @ Signal_EXMEM_FA_Mux[0] @
                                        Signal_EXMEM_FA_Mux[3:1];
            }
            else
            {delay;}
        }
        else if(Signal_EXMEM_ALU_Ctrl == 0b1101)  // FREE
        {
            delay;
        }
        else if(Signal_EXMEM_ALU_Ctrl == 0b1110)  // FREE
        {
            delay;
        }
        else if(Signal_EXMEM_ALU_Ctrl == 0b1111)  // FREE
        {
            delay;
        }
        else
        {delay;}

        // Destination register selection from either rd(R-FORMAT) or rt
            (I-FORMAT)
        if(Signal_EXMEM_RegDst == 1)
        {Signal_EXMEM_Regd = Signal_EXMEM_Rd;}
        else
        {Signal_EXMEM_Regd = Signal_EXMEM_Rt;}

        // Memory Read / Write
        if((Signal_EXMEM_MemRead == 1) && (Signal_EXMEM_MemWrite != 1))
```

```
{
    Signal_EXMEM_MemData = 0@STRIP_MEMORY[Signal_EXMEM_FA_Mux
        [12:0]];
}
else
{delay;}

if((Signal_EXMEM_MemRead != 1) && (Signal_EXMEM_MemWrite == 1))
{
    STRIP_MEMORY[Signal_EXMEM_FA_Mux[12:0]] =
        Signal_EXMEM_FB_Mux[15:0];
}
else
{delay;}

// Stage 4: Write Back ----------------------------------------
Signal_WB_RegWrite         = EXMEMWB_WB_RegWrite;
Signal_WB_Rd               = EXMEMWB_Rd;
Signal_WB_MemtoReg         = EXMEMWB_WB_MemtoReg;
Signal_WB_ALUData          = EXMEMWB_WB_ALUData;
Signal_WB_MemData          = EXMEMWB_WB_MemData;

if(Signal_WB_MemtoReg == 1)
{Signal_WB_Data = Signal_WB_MemData;}
else
{Signal_WB_Data = Signal_WB_ALUData;}

//------------------------------------------------------------
// New clock cycle
//------------------------------------------------------------

// Stage 1: Instruction Fetch ---------------------------------
if(Signal_ID_PCWrite == 1)
{PC = Signal_IF_PC_Mux;}
else{delay;}

if(Signal_ID_IFIDWrite == 1)
{IFID_Inst      = Signal_IF_Inst;}
else{delay;}

// Stage 2: Instruction Decode --------------------------------
IDEXMEM_WB_MemtoReg    = Signal_ID_Hzd_MemtoReg;
IDEXMEM_WB_RegWrite    = Signal_ID_Hzd_RegWrite;
IDEXMEM_MEM_MemRead    = Signal_ID_Hzd_MemRead;
IDEXMEM_MEM_MemWrite   = Signal_ID_Hzd_MemWrite;
IDEXMEM_EX_RegDst      = Signal_ID_Hzd_RegDst;
IDEXMEM_EX_ALUOp1      = Signal_ID_Hzd_ALUOp1;
IDEXMEM_EX_ALUOp0      = Signal_ID_Hzd_ALUOp0;
IDEXMEM_EX_ALUSrc      = Signal_ID_Hzd_ALUSrc;

IDEXMEM_ReadData1      = Signal_ID_ReadData1;
IDEXMEM_ReadData2      = Signal_ID_ReadData2;
IDEXMEM_ImmVal         = Signal_ID_Inst[15:0];
IDEXMEM_Rs             = Signal_ID_Hzd_Rs;
IDEXMEM_Rt             = Signal_ID_Hzd_Rt;
IDEXMEM_Rd             = Signal_ID_Hzd_Rd;

// Stage 3: Execution and Memory Access -----------------------
EXMEMWB_WB_RegWrite        = Signal_EXMEM_RegWrite;
EXMEMWB_WB_MemtoReg        = Signal_EXMEM_MemtoReg;
EXMEMWB_Rd                 = Signal_EXMEM_Regd;
EXMEMWB_WB_ALUData         = Signal_EXMEM_ALU_Result;
EXMEMWB_WB_MemData         = Signal_EXMEM_MemData;
```

```
            //-------------------------------------------------------------
            // Update counter / termination condition
            //-------------------------------------------------------------
            ClockCycle = ClockCycle + 1;      // increment clock cycle
            if(PC == InstMemSize)             // if reached the last
                instruction
            {
                RUN = 0;                      // set termination condition
            }
            else {delay;}
        } // end par
    } // end while
}

static macro proc OutputStripMemory()
{
    unsigned int PatchCountBits       PatchCounter;
    unsigned int PatchPixelCountBits  PatchPixelCounter, RCounter;
    unsigned int PatchWidthCountBits  PatchWidthCounter;
    unsigned int StripBits            WriteData;
    unsigned int RAMStripAddressBits  StartStripAddressPatchData;
    unsigned int RAMStripAddressBits  StartStripAddressPatchHeader;
    unsigned int PatchCountBits       RowPatchCounter;
    unsigned int PatchCountBits       ColPatchCounter;
    unsigned int PatchCountBits       RAMAddressCounter;

    // Zigbee transmission variables - using API packet
    unsigned 8 StartDelimiter, FrameType, FrameID, BroadcastRadius, Options;
    unsigned 8 DestAddr64_0, DestAddr64_1, DestAddr64_2, DestAddr64_3,
        DestAddr64_4, DestAddr64_5, DestAddr64_6, DestAddr64_7;
    unsigned 8 DestAddr16_MSB, DestAddr16_LSB;
    unsigned 16 Length, NumofPatches, TempCheck, CheckSum, PatchWidthNum;

    seq
    {
        // Parameter settings which cannot be assigned to unsigned from
            macro expr goes here
        NumofPatches = 32; // number of patches used per strip image
        PatchWidthNum = 8;

        // First send data containing valid patches information (saliency
            bit = 1 or 0)
        StartDelimiter = 0x7E;
        Length = 14 + NumofPatches;
        FrameType = 0x10;
        FrameID = 0x01;
        DestAddr64_0 = 0x00;
        DestAddr64_1 = 0x00;
        DestAddr64_2 = 0x00;
        DestAddr64_3 = 0x00;
        DestAddr64_4 = 0x00;
        DestAddr64_5 = 0x00;
        DestAddr64_6 = 0x00;
        DestAddr64_7 = 0x00;
        DestAddr16_MSB = 0xFF;
        DestAddr16_LSB = 0xFE;
        BroadcastRadius = 0x00;
        Options = 0x00;

        TempCheck = 0;
        TempCheck = 0@(FrameType + FrameID + DestAddr64_0 + DestAddr64_1 +
            DestAddr64_2 + DestAddr64_3 + DestAddr64_4 + DestAddr64_5 +
            DestAddr64_6 + DestAddr64_7 + DestAddr16_MSB + DestAddr16_LSB +
            BroadcastRadius + Options);

        // Send API packet header
        RC10RS232Write(StartDelimiter);
```

```
RC10RS232Write(Length[15:8]);
RC10RS232Write(Length[7:0]);
RC10RS232Write(FrameType);
RC10RS232Write(FrameID);
RC10RS232Write(DestAddr64_0);
RC10RS232Write(DestAddr64_1);
RC10RS232Write(DestAddr64_2);
RC10RS232Write(DestAddr64_3);
RC10RS232Write(DestAddr64_4);
RC10RS232Write(DestAddr64_5);
RC10RS232Write(DestAddr64_6);
RC10RS232Write(DestAddr64_7);
RC10RS232Write(DestAddr16_MSB);
RC10RS232Write(DestAddr16_LSB);
RC10RS232Write(BroadcastRadius);
RC10RS232Write(Options);

// Send API Data
RAMAddressStripMemory = RAMStripAddress - NumPatches;
PatchCounter = 0;
while (PatchCounter != NumPatches)
{
    seq
    {
        WriteData = STRIP_MEMORY[RAMAddressStripMemory];
        RC10RS232Write(WriteData[7:0]);
        TempCheck = TempCheck + (0@(WriteData[7:0]));
        RAMAddressStripMemory++;
        PatchCounter++;
    }
}

// Send API checksum
CheckSum = 0xFF - TempCheck;
RC10RS232Write(CheckSum[7:0]);

// Next transmit image data

StartStripAddressPatchHeader = RAMStripAddress - NumPatches;
StartStripAddressPatchData   = 0;
RowPatchCounter = 0;
ColPatchCounter = 0;

while(RowPatchCounter != PatchesPerRow)
{
    seq
    {
        while(ColPatchCounter != PatchesPerCol)
        {
            seq
            {
                RAMAddressStripMemory = StartStripAddressPatchHeader
                    ;
                if(STRIP_MEMORY[RAMAddressStripMemory] == 1)
                {
                    seq
                    {
                        RAMAddressStripMemory =
                            StartStripAddressPatchData;
                        PatchWidthCounter = 0;
                        PatchPixelCounter = 0;

                        StartDelimiter = 0x7E;
                        Length = 14 + 64;
                        FrameType = 0x10;
                        FrameID = 0x01;
                        DestAddr64_0 = 0x00;
```

```
                        DestAddr64_1 = 0x00;
                        DestAddr64_2 = 0x00;
                        DestAddr64_3 = 0x00;
                        DestAddr64_4 = 0x00;
                        DestAddr64_5 = 0x00;
                        DestAddr64_6 = 0x00;
                        DestAddr64_7 = 0x00;
                        DestAddr16_MSB = 0xFF;
                        DestAddr16_LSB = 0xFE;
                        BroadcastRadius = 0x00;
                        Options = 0x00;

                        TempCheck = 0;
                        TempCheck = 0@(FrameType + FrameID +
                            DestAddr64_0 + DestAddr64_1 +
                            DestAddr64_2 + DestAddr64_3 +
                            DestAddr64_4 + DestAddr64_5 +
                            DestAddr64_6 + DestAddr64_7 +
                            DestAddr16_MSB + DestAddr16_LSB +
                            BroadcastRadius + Options);

                        // Send API packet header
                        RC10RS232Write(StartDelimiter);
                        RC10RS232Write(Length[15:8]);
                        RC10RS232Write(Length[7:0]);
                        RC10RS232Write(FrameType);
                        RC10RS232Write(FrameID);
                        RC10RS232Write(DestAddr64_0);
                        RC10RS232Write(DestAddr64_1);
                        RC10RS232Write(DestAddr64_2);
                        RC10RS232Write(DestAddr64_3);
                        RC10RS232Write(DestAddr64_4);
                        RC10RS232Write(DestAddr64_5);
                        RC10RS232Write(DestAddr64_6);
                        RC10RS232Write(DestAddr64_7);
                        RC10RS232Write(DestAddr16_MSB);
                        RC10RS232Write(DestAddr16_LSB);
                        RC10RS232Write(BroadcastRadius);
                        RC10RS232Write(Options);

                        while(PatchPixelCounter != (PatchWidth*
                            PatchHeight))
                        {
                            seq
                            {
                                WriteData = STRIP_MEMORY[
                                    RAMAddressStripMemory];
                                RC10RS232Write(WriteData[7:0]);
                                TempCheck = TempCheck + (0@(
                                    WriteData[7:0]));
                                if(PatchWidthCounter == (PatchWidth
                                    - 1))
                                {
                                    par
                                    {
                                        PatchWidthCounter      = 0;
                                        RAMAddressStripMemory =
                                            RAMAddressStripMemory +
                                            (ImageWidth -
                                            PatchWidth + 1);
                                    }
                                }
                                else
                                {
                                    par
                                    {
                                        PatchWidthCounter++;
```

```
                                        RAMAddressStripMemory++;
                               }
                       }

                       PatchPixelCounter++;
               }
           }
           // Send API checksum
           CheckSum = 0xFF - TempCheck;
           RC10RS232Write(CheckSum[7:0]);
       }
   }

   StartStripAddressPatchHeader++;
   StartStripAddressPatchData =
       StartStripAddressPatchData + PatchWidth;
   ColPatchCounter++;
       }
   }
   RAMAddressCounter = 0;
   while(RAMAddressCounter != PatchHeight - 1)
   {
       seq
       {
       StartStripAddressPatchData =
           StartStripAddressPatchData + ImageWidth;
       RAMAddressCounter++;
       }
   }
   RowPatchCounter++;
   ColPatchCounter = 0;
       }
   }

       }
   }
```

Listing A.1 Handel-C Program Listing for WMSN FPGA Processor

References

1. Achanta, R., Estrada, F., Wils, P., Süsstrunk, S.: Salient region detection and segmentation. In: Proceedings of the 6th International Conference on Computer Vision Systems (ICVS'08), pp. 66–75. Springer, Berlin (2008). URL http://dl.acm.org/citation.cfm?id=1788524.1788532
2. Achanta, R., Hemami, S., Estrada, F., Susstrunk, S.: Frequency-tuned salient region detection. In: IEEE Conference on Computer Vision and Pattern Recognition, 2009 (CVPR 2009), pp. 1597–1604 (2009). doi:10.1109/CVPR.2009.5206596
3. Acharya, T., Tsai, P.S.: JPEG2000 Standard for Image Compression: Concepts, Algorithms and VLSI Architectures. Wiley-Interscience, Hoboken (2004)
4. Aghajan, H., Cavallaro, A.: Multi-Camera Networks: Principles and Applications. Academic, London (2009)
5. Akyildiz, I.F., Melodia, T., Chowdhury, K.R.: A survey on wireless multimedia sensor networks. Comput. Netw. **51**(4), 921–960 (2007). doi:10.1016/j.comnet.2006.10.002. URL http://dx.doi.org/10.1016/j.comnet.2006.10.002
6. Almalkawi, I., Zapata, M., Al-Karaki, J., Morillo-Pozo, J.: Wireless multimedia sensor networks: Current trends and future directions. Sensors (Basel) **10**(7), 6662–6717 (2010)
7. Altera: Altera Stratix-V. http://www.altera.com/devices/fpga/stratix-fpgas/stratix-v/stxv-index.jsp (2013)
8. Altera: Nios II Processor. http://www.altera.com.my/devices/processor/nios2/ni2-index.html (2011)
9. Altera: Stratix II datasheet. http://www.altera.com.my/literature/hb/stx2/stx2_sii51002.pdf (2007)
10. Ammari, A., Jemai, A.: Multiprocessor platform-based design for multimedia. IET Comput. Digit. Tech. **3**(1), 52–61 (2009). doi:10.1049/iet-cdt:20070168
11. Angelopoulou, M.E., Masselos, K., Cheung, P.Y., Andreopoulos, Y.: Implementation and comparison of the 5/3 lifting 2d discrete wavelet transform computation schedules on fpgas. J. Signal Process. Syst. **51**(1), 3–21 (2008). doi:10.1007/s11265-007-0139-5. URL http://dx.doi.org/10.1007/s11265-007-0139-5
12. ARM Ltd.: ARM architectures. http://www.arm.com/products/processors/index.php (2011)
13. Arnold, M.G.: Verilog Digital Computer Design: Algorithms into Hardware. Prentice Hall, Upper Saddle River (1999)
14. Bhattar, R.K., Ramakrishnan, K., Dasgupta, K.: Strip based coding for large images using wavelets. Signal Process. Image Commun. **17**(6), 441–456 (2002). doi: 10.1016/S0923-5965(02)00019-X. URL http://www.sciencedirect.com/science/article/pii/S092359650200019X
15. Brown, M., Lowe, D.G.: Automatic panoramic image stitching using invariant features. Int. J. Comput. Vis. **74**(1), 59–73 (2007). doi:10.1007/s11263-006-0002-3. URL http://dx.doi.org/10.1007/s11263-006-0002-3

L. Ang et al., *Wireless Multimedia Sensor Networks on Reconfigurable Hardware*,
DOI 10.1007/978-3-642-38203-1, © Springer-Verlag Berlin Heidelberg 2013

16. Bruce, N.D.B., Tsotsos, J.K.: Saliency, attention, and visual search: An information theoretic approach. J. Vis. **9**(3), 1–24 (2009)

17. Burt, P.J., Adelson, E.H.: A multiresolution spline with application to image mosaics. ACM Trans. Graph. **2**(4), 217–236 (1983). doi:10.1145/245.247. URL http://doi.acm.org/10.1145/245.247

18. Cadambi, S., Weener, J., Goldstein, S.C., Schmit, H., Thomas, D.E.: Managing pipeline-reconfigurable fpgas. In: Proceedings of the 1998 ACM/SIGDA Sixth International Symposium on Field Programmable Gate Arrays (FPGA '98), pp. 55–64. ACM, New York (1998). doi:10.1145/275107.275120. URL http://doi.acm.org/10.1145/275107.275120

19. Cardoso, J.F.: High-order contrasts for independent component analysis. Neural Comput. **11**, 157–192 (1999). doi:http://dx.doi.org/10.1162/089976699300016863. URL http://dx.doi.org/10.1162/089976699300016863

20. Cassereau, P.M.: Wavelet-based image coding. In: Watson, A.B. (ed.) Digital Images and Human Vision, pp. 12–21. The MIT Press, Cambridge (1993)

21. Celoxica: Celoxica RC10. www.europractice.rl.ac.uk/vendors/agility_rc10.pdf (2005)

22. Celoxica: Celoxica RC230E. http://babbage.cs.qc.edu/courses/cs345/Manuals/RC200_203%20Manual.pdf (2005)

23. Celoxica: DK Design Suite. http://www.europractice.stfc.ac.uk/vendors/agility_dk4.pdf (2005)

24. Celoxica: DK4 Handel-C Language Reference Manual. http://babbage.cs.qc.edu/courses/cs345/Manuals/HandelC.pdf (2005)

25. Charfi, Y., Wakamiya, N., Murata, M.: Network-adaptive image and video transmission in camera-based wireless sensor networks. In: IEEE International Conference on Distributed Smart Cameras, pp. 336–343 (2007). URL http://dblp.uni-trier.de/db/conf/icdsc/icdsc2007.html#CharfiWM07

26. Cheung, S.C.S., Kamath, C.: Robust background subtraction with foreground validation for urban traffic video. EURASIP J. Appl. Signal Process. **14**, 2330–2340 (2005). doi:10.1155/ASP.2005.2330. URL http://dx.doi.org/10.1155/ASP.2005.2330

27. Chew, L.W., Ang, L.M., Seng, K.P.: New virtual spiht tree structures for very low memory strip-based image compression. IEEE Signal Process. Lett. **15**, 389 –392 (2008). doi:10.1109/LSP.2008.920515

28. Chew, L.W., Chia, W.C., Ang, L.M., Seng, K.P.: Very low-memory wavelet compression architecture using strip-based processing for implementation in wireless sensor networks. EURASIP J. Embed. Syst. **2009**, 9:1–9:1 (2009). doi:10.1155/2009/479281. URL http://dx.doi.org/10.1155/2009/479281

29. Chew, L.W., Chia, W.C., Ang, L.M., Seng, K.P.: Low-memory video compression architecture using strip-based processing for implementation in wireless multimedia sensor networks. Int. J. Sens. Netw. **11**(1), 33–47 (2012). doi:10.1504/IJSNET.2012.045033. URL http://dx.doi.org/10.1504/IJSNET.2012.045033

30. Chia, W.C., Chew, L.W., Ang, L.M., Seng, K.P.: Low memory image stitching and compression for wmsn using strip-based processing. IJSNet **11**(1), 22–32 (2012). URL http://dblp.uni-trier.de/db/journals/ijsnet/ijsnet11.html#ChiaCAS12

31. Choi, S., Scrofano, R., Prasanna, V.K., Jang, J.W.: Energy-efficient signal processing using FPGAs. In: Proceedings of the 2003 ACM/SIGDA Eleventh International Symposium on Field Programmable Gate Arrays (FPGA '03), pp. 225–234. ACM, New York (2003). doi:10.1145/611817.611850. URL http://doi.acm.org/10.1145/611817.611850

32. Chrysafis, C., Ortega, A.: Line-based, reduced memory, wavelet image compression. IEEE Trans. Image Process. **9**(3), 378–389 (2000). doi:10.1109/83.826776

33. Compton, K., Hauck, S.: Reconfigurable computing: A survey of systems and software. ACM Comput. Surv. **34**(2), 171–210 (2002). doi:10.1145/508352.508353. URL http://doi.acm.org/10.1145/508352.508353

34. Crossbow Technology: XBow SPB400–Stargate Gateway Datasheet. http://bullseye.xbow.com:81/Products/Product_pdf_files/Wireless_pdf/Stargate_Datasheet.pdf (2006)

35. Crossbow Technology: XBow TELOSB Datasheet. http://bullseye.xbow.com:81/Products/Product_pdf_files/Wireless_pdf/TelosB_Datasheet.pdf (2007)
36. Cucchiara, R.: Multimedia surveillance systems. In: Proceedings of the Third ACM International Workshop on Video Surveillance & Sensor Networks (VSSN '05), pp. 3–10. ACM, New York (2005). doi:10.1145/1099396.1099399. URL http://doi.acm.org/10.1145/1099396.1099399
37. Czarlinska, A., Kundur, D.: Reliable event-detection in wireless visual sensor networks through scalar collaboration and game-theoretic consideration. IEEE Trans. Multimed. **10**(5), 675–690 (2008). doi:10.1109/TMM.2008.922775
38. Daubechies, I., Sweldens, W.: Factoring wavelet transforms into lifting steps. J. Fourier Anal. Appl. **4**(3), 247–269 (1998)
39. Elazary, L., Itti, L.: A bayesian model for efficient visual search and recognition. Vis. Res. **50**(14), 1338–1352 (2010)
40. Estrin, G.: Organization of computer systems: The fixed plus variable structure computer. In: Papers Presented at the May 3–5, 1960, Western Joint IRE-AIEE-ACM Computer Conference (IRE-AIEE-ACM '60) (Western), pp. 33–40. ACM, New York (1960). doi:10.1145/1460361.1460365. URL http://doi.acm.org/10.1145/1460361.1460365
41. Faraji, I., Didehban, M., Zarandi, H.: Analysis of transient faults on a mips-based dual-core processor. In: International Conference on Availability, Reliability, and Security, 2010 (ARES '10), pp. 125–130 (2010). doi:10.1109/ARES.2010.30
42. Fischler, M.A., Bolles, R.C.: Random sample consensus: A paradigm for model fitting with applications to image analysis and automated cartography. Commun. ACM **24**(6), 381–395 (1981). doi:10.1145/358669.358692. URL http://doi.acm.org/10.1145/358669.358692
43. Foulsham, T., Underwood, G.: What can saliency models predict about eye movements? spatial and sequential aspects of fixations during encoding and recognition. J. Vis. **8**(2) (2008). doi:10.1167/8.2.6. URL http://www.journalofvision.org/content/8/2/6.abstract
44. Frintrop, S.: Computational visual attention. In: Computer Analysis of Human Behavior, pp. 69–101. Springer, London (2011)
45. Frintrop, S., Jensfelt, P.: Attentional landmarks and active gaze control for visual slam. IEEE Trans. Robot. **24**(5), 1054–1065 (2008). doi:10.1109/TRO.2008.2004977
46. Fry, T., Hauck, S.: SPIHT image compression on FPGAs. IEEE Trans. Circuits Syst. Video Technol. **15**(9), 1138–1147 (2005). doi:10.1109/TCSVT.2005.852625
47. Garcia, V.: Keypoints Extraction. http://www.mathworks.com/matlabcentral/fileexchange/17894-keypoint-extraction (2007)
48. Gautham, P., Parthasarathy, R., Balasubramanian, K.: Low-power pipelined mips processor design. In: Proceedings of the 2009 12th International Symposium on Integrated Circuits (ISIC '09), pp. 462–465 (2009)
49. Girod, B., Aaron, A., Rane, S., Rebollo-Monedero, D.: Distributed video coding. Proc. IEEE **93**(1), 71 –83 (2005). doi:10.1109/JPROC.2004.839619
50. Gray, A., Lee, C., Arabshahi, P., Srinivasan, J.: Object-oriented reconfigurable processing for wireless networks. In: IEEE International Conference on Communications, 2002 (ICC 2002), vol. 1, pp. 497–501 (2002). doi:10.1109/ICC.2002.996903
51. Guo, C., Zhang, L.: A novel multiresolution spatiotemporal saliency detection model and its applications in image and video compression. IEEE Trans. Image Process. **19**(1), 185–198 (2010). doi:10.1109/TIP.2009.2030969
52. Guo, W., Xu, C., Ma, S., Xu, M.: Visual attention based small object segmentation in natual images. In: 2010 17th IEEE International Conference on Image Processing (ICIP), pp. 1565–1568 (2010). doi:10.1109/ICIP.2010.5649841
53. Guo, X., Lu, Y., Wu, F., Zhao, D., Gao, W.: Wyner–ziv-based multiview video coding. IEEE Trans. Circuits Syst. Video Technol. **18**(6), 713–724 (2008). doi:10.1109/TCSVT.2008.920970. URL http://dx.doi.org/10.1109/TCSVT.2008.920970
54. Gürses, E., Akan, Ö.: Multimedia communication in wireless sensor networks. Ann. Telecommun. **60**(7), 872–900 (2005)

55. Haering, N., da Vitoria Lobo, N.: Visual Event Detection. The International Series in Video Computing. Springer, New York (2001). URL http://www.springer.com/computer/image+processing/book/978-0-7923-7436-7

56. Harris, C., Stephens, M.: A combined corner and edge detector. In: Proceedings of Fourth Alvey Vision Conference, pp. 147–151 (1988)

57. HART Communication Protocol and Foundation: WirelessHART Overview. http://www.hartcomm.org/protocol/wihart/wireless_overview.html (2010)

58. Hasan, M., Rahman, M., Hasan, M., Hasan, M., Ali, M.: An improved pipelined processor architecture eliminating branch and jump penalty. In: 2010 Second International Conference on Computer Engineering and Applications (ICCEA), vol. 1, pp. 621–625 (2010). doi: 10.1109/ICCEA.2010.126

59. Hauser, J., Wawrzynek, J.: Garp: A mips processor with a reconfigurable coprocessor. In: Proceedings of the 5th Annual IEEE Symposium on Field-Programmable Custom Computing Machines, 1997, pp. 12–21 (1997). doi:10.1109/FPGA.1997.624600

60. Hill, J., Szewczyk, R., Woo, A., Hollar, S., Culler, D., Pister, K.: System architecture directions for networked sensors. SIGPLAN Not. **35**(11), 93–104 (2000). doi: 10.1145/356989.356998. URL http://doi.acm.org/10.1145/356989.356998

61. Ho-Phuoc, T., Guyader, N., Gurin-Dugu, A.: A functional and statistical bottom-up saliency model to reveal the relative contributions of low-level visual guiding factors. Cogn. Comput. **2**, 344–359 (2010). doi:10.1007/s12559-010-9078-8. URL http://dx.doi.org/10.1007/s12559-010-9078-8

62. Hu, F., Kumar, S.: QoS considerations in wireless sensor networks for telemedicine. In: Proceedings of SPIE ITCOM Conference, Orlando, FL (2003)

63. Hunter, R.D.M., Johnson, T.T.: Introduction to VHDL. Springer, Berlin (1995). http://www.springer.com/engineering/circuits+%26+systems/book/978-0-412-73130-3

64. International, D.: XBee ZNet 2.5 Module. http://www.digi.com/support/productdetail?pid=3261 (2012). Accessed Nov 2012

65. International, D.: XBee/XBee-Pro RF Module Datasheet. http://www.digi.com/pdf/ds_xbeezbmodules.pdf (2011). Accessed Nov 2012

66. Itti, L., Koch, C.: A saliency-based search mechanism for overt and covert shifts of visual attention. Vis. Res. **40**, 1489–1506 (2000)

67. Itti, L., Koch, C., Niebur, E.: A model of saliency-based visual attention for rapid scene analysis. IEEE Trans. Pattern Anal. Mach. Intell. **20**(11), 1254–1259 (1998). doi:10.1109/34.730558

68. Itti, L., Rees, G., Tsotsos, J.: Models of bottom-up attention and saliency. In: Neurobiology of Attention, pp. 576–582. Elsevier, San Diego (2005)

69. James, W.: The Principles of Psychology. American Science Series: Advanced Course, vol. 1. H. Holt, New York (1918). URL http://books.google.com.my/books?id=lbtE-xb5U-oC

70. Kahn, J.M., Katz, R.H., Pister, K.S.J.: Next century challenges: Mobile networking for Smart Dust. In: Proceedings of the 5th Annual ACM/IEEE International Conference on Mobile Computing and Networking (MobiCom '99), pp. 271–278. ACM, New York (1999). doi:10.1145/313451.313558. URL http://doi.acm.org/10.1145/313451.313558

71. Koch, C., Ullman, S.: Shifts in selective visual attention: Towards the underlying neural circuitry. Hum. Neurobiol. **4**(4), 219–227 (1985)

72. Kunt, M., Ikonomopoulos, A., Kocher, M.: Second-generation image-coding techniques. Proc. IEEE **73**(4), 549–574 (1985). doi:10.1109/PROC.1985.13184

73. Kwok, T.T.O., Kwok, Y.K.: Computation and energy efficient image processing in wireless sensor networks based on reconfigurable computing. In: Proceedings of the 2006 International Conference Workshops on Parallel Processing (ICPPW '06), pp. 43–50. IEEE Computer Society, Washington, DC (2006). doi:10.1109/ICPPW.2006.30. URL http://dx.doi.org/10.1109/ICPPW.2006.30

74. Lee, S.H., Shin, J.K., Lee, M.: Non-uniform image compression using biologically motivated saliency map model. In: Proceedings of the 2004 Intelligent Sensors, Sensor

Networks and Information Processing Conference, 2004, pp. 525–530 (2004)   doi: 10.1109/ISSNIP.2004.1417516

75. Li, L.J., Su, H., Xing, E.P., Fei-Fei, L.: Object bank: A high-level image representation for scene classification and semantic feature sparsification. In: Advances in Neural Information Processing Systems. MIT Press, Cambridge (2010)

76. Li, X., Li, T.: Ecomips: An economic MIPS CPU design on FPGA. In: Proceedings of the 4th IEEE International Workshop on System-on-Chip for Real-Time Applications, 2004, pp. 291–294 (2004). doi:10.1109/IWSOC.2004.1319896

77. Liang, J., DeMenthon, D., Doermann, D.: Note: Mosaicing of camera-captured document images. Comput. Vis. Image Underst. 113(4), 572–579 (2009).   doi: 10.1016/j.cviu.2008.12.004. URL http://dx.doi.org/10.1016/j.cviu.2008.12.004

78. Libelium Comunicaciones Distribuidas S.L.: Libelium Waspmote technical guide. http:// www.libelium.com/downloads/documentation/waspmote_technical_guide.pdf (2013)

79. Lindeberg, T.: Feature detection with automatic scale selection. Int. J. Comput. Vis. 30(2), 79–116 (1998). doi:10.1023/A:1008045108935. URL http://dx.doi.org/10.1023/A: 1008045108935

80. Liu, L., Yuan, F.G., Zhang, F.: Development of wireless smart sensor for structural health monitoring. In: Proceedings of the 2005 Smart Structures and Materials - Sensors and Smart Structures Technologies for Civil, Mechanical, and Aerospace Systems, vol. 5765, No. 20, pp. 176–186 (2005). doi:10.1117/12.600206. URL http://dx.doi.org/10.1117/12.600206

81. Liu, T., Yuan, Z., Sun, J., Wang, J., Zheng, N., Tang, X., Shum, H.Y.: Learning to detect a salient object. IEEE Trans. Pattern Anal. Mach. Intell. 33(2), 353 –367 (2011). doi:10.1109/TPAMI.2010.70

82. Liu, Y., Xie, G., Chen, P., Chen, J., Li, Z.: Designing an asynchronous FPGA processor for low-power sensor networks. In: International Symposium on Signals, Circuits and Systems, 2009 (ISSCS 2009), pp. 1–6 (2009). doi:10.1109/ISSCS.2009.5206091

83. Lowe, D.G.: Distinctive image features from scale-invariant keypoints. Int. J. Comput. Vis. 60(2), 91–110 (2004). doi:10.1023/B:VISI.0000029664.99615.94. URL http://dx.doi.org/ 10.1023/B:VISI.0000029664.99615.94

84. Lukai Cai, S.V., Gajski, D.D.: Technical report cecs-03-11. Technical Report, Center for Embedded Computer Systems, University of California, Irvine, California, 2003

85. Magli, E., Mancin, M., Merello, L.: Low-complexity video compression for wireless sensor networks. In: Proceedings of the 2003 International Conference on Multimedia and Expo - Vol. 3 (ICME '03), pp. 585–588. IEEE Computer Society, Washington, DC (2003). URL http://dl.acm.org/citation.cfm?id=1170746.1171773

86. Mangharam, R., Rowe, A., Rajkumar, R.: FireFly: A cross-layer platform for real-time embedded wireless networks. R. Time Syst. 37(3), 183–231 (2007). doi: 10.1007/s11241-007-9028-z. URL http://dx.doi.org/10.1007/s11241-007-9028-z

87. Martin, D., Fowlkes, C., Tal, D., Malik, J.: The berkeley segmentation dataset. http://www. eecs.berkeley.edu/Research/Projects/CS/vision/bsds/ (2007)

88. Medeiros, H., Park, J., Kak, A.: Distributed object tracking using a cluster-based kalman filter in wireless camera networks. IEEE J. Sel. Top. Signal Process. 2(4), 448–463 (2008). doi:10.1109/JSTSP.2008.2001310

89. Mendi, E., Milanova, M.: Image segmentation with active contours based on selective visual attention. In: Proceedings of the 3rd WSEAS International Symposium on Wavelets Theory and Applications in Applied Mathematics, Signal Processing & Modern Science (WAV'09), pp. 79–84. World Scientific and Engineering Academy and Society (WSEAS), Stevens Point, Wisconsin, USA (2009). URL http://dl.acm.org/citation.cfm?id=1561963.1561976

90. (MERL), M.E.R.L.: Stereoscopic Video Sequences. ftp://ftp.merl.com/pub/avetro/mvc-testseq/orig-yuv/exit/ (2005)

91. Merrill, W., Sohrabi, K., Girod, L., Elson, J., Newberg, F.: Open standard development platforms for distributed sensor networks. In: SPIE Unattended Ground Sensor Technologies and Applications IV, pp. 327–337 (2002)

92. Mikolajczyk, K., Schmid, C.: Scale and affine invariant interest point detectors. Int. J. Comput. Vis. **60**(1), 63–86 (2004). doi:10.1023/B:VISI.0000027790.02288.f2. URL http://dx.doi.org/10.1023/B:VISI.0000027790.02288.f2

93. Mikolajczyk, K., Schmid, C.: A performance evaluation of local descriptors. IEEE Trans. Pattern Anal. Mach. Intell. **27**(10), 1615–1630 (2005). doi:10.1109/TPAMI.2005.188. URL http://dx.doi.org/10.1109/TPAMI.2005.188

94. Mills, A., Dudek, G.: Image stitching with dynamic elements. Image Vis. Comput. **27**(10), 1593–1602 (2009). doi:10.1016/j.imavis.2009.03.004. URL http://dx.doi.org/10.1016/j.imavis.2009.03.004

95. MIPS Technologies: MIPS architectures. http://www.mips.com/products/architectures/mips32/ (2008)

96. Mulligan, G.: The 6lowpan architecture. In: Proceedings of the 4th Workshop on Embedded Networked Sensors (EmNets '07), pp. 78–82. ACM, New York (2007). doi:10.1145/1278972.1278992. URL http://doi.acm.org/10.1145/1278972.1278992

97. Ngau, C., Ang, L.M., Seng, K.: Low memory visual saliency architecture for data reduction in wireless sensor networks. IET Wirel. Sens. Syst. **2**(2), 115 –127 (2012). doi:10.1049/iet-wss.2011.0038

98. Omnivision: Omnivision OV16820. http://www.ovt.com/products/sensor.php?id=116 (2012)

99. Ouerhani, N., Bracamonte, J., Hugli, H., Ansorge, M., Pellandini, F.: Adaptive color image compression based on visual attention. In: Proceedings of the 11th International Conference on Image Analysis and Processing, 2001, pp. 416–421 (2001). doi:10.1109/ICIAP.2001.957045

100. Patterson, D.A., Hennessy, J.L.: Computer Organization and Design: The Hardware/Software Interface, 3rd edn. Morgan Kaufmann Publishers Inc., San Francisco (2007)

101. Pearlman, W.A., Islam, A., Nagaraj, N., Said, A.: Efficient, low-complexity image coding with a set-partitioning embedded block coder. IEEE Trans. Circuits Syst. Video Technol. **14**(11), 1219–1235 (2004). doi:10.1109/TCSVT.2004.835150. URL http://dx.doi.org/10.1109/TCSVT.2004.835150

102. Pham, D.M., Aziz, S.: FPGA-based image processor architecture for wireless multimedia sensor network. In: 2011 IFIP 9th International Conference on Embedded and Ubiquitous Computing (EUC), pp. 100–105 (2011). doi:10.1109/EUC.2011.38

103. Pradhan, S.S., Ramchandran, K.: Distributed source coding using syndromes (discus): Design and construction. IEEE Trans. Inf. Theory **49**(3), 626–643 (2006). doi:10.1109/TIT.2002.808103. URL http://dx.doi.org/10.1109/TIT.2002.808103

104. Puri, R., Ramchandran, K.: Prism: An uplink-friendly multimedia coding paradigm. In: Proceedings of the 2003 IEEE International Conference on Acoustics, Speech, and Signal Processing, 2003 (ICASSP '03), vol. 4, pp. IV–856–859 (2003). doi:10.1109/ICASSP.2003.1202778

105. Quattoni, A., Torralba, A.: Indoor scene recognition database. http://web.mit.edu/torralba/www/indoor.html (2009)

106. Rajagopalan, R., Varshney, P.: Data-aggregation techniques in sensor networks: a survey. IEEE Commun. Surv. Tutor. **8**(4), 48 –63 (2006). doi:10.1109/COMST.2006.283821

107. Ramdas, T., Ang, L.M., Egan, G.: Fpga implementation of an integer mips processor in handel-c and its application to human face detection. In: TENCON 2004. 2004 IEEE Region 10 Conference Volume A, vol. 1, pp. 36–39 (2004). doi:10.1109/TENCON.2004.1414350

108. Rapantzikos, K., Avrithis, Y., Kollias, S.: Spatiotemporal saliency for event detection and representation in the 3d wavelet domain: Potential in human action recognition. In: Proceedings of the 6th ACM International Conference on Image and Video Retrieval (CIVR '07), pp. 294–301. ACM, New York (2007). doi:10.1145/1282280.1282326. URL http://doi.acm.org/10.1145/1282280.1282326

109. Reid, M.M., Millar, R.J., Black, N.D.: Second-generation image coding: An overview. ACM Comput. Surv. **29**(1), 3–29 (1997). doi:10.1145/248621.248622. URL http://doi.acm.org/10.1145/248621.248622

110. Rosten, E., Drummond, T.: Machine learning for high-speed corner detection. In: Proceedings of the 9th European Conference on Computer Vision - Volume Part I (ECCV'06), pp. 430–443. Springer, Berlin, Heidelberg (2006). doi:10.1007/11744023_34. URL http://dx.doi.org/10.1007/11744023_34

111. Said, A., Pearlman, W.A.: A new, fast, and efficient image codec based on set partitioning in hierarchical trees. IEEE Trans. Circuits Syst. Video Technol. 6(3), 243–250 (1996). doi:10.1109/76.499834. URL http://dx.doi.org/10.1109/76.499834

112. Sensirion AG: Sensirion Digital Humidity and Temperature Sensors (RH&T). http://www.sensirion.com/en/products/humidity-temperature/ (2011)

113. Shapiro, J.: Embedded image coding using zerotrees of wavelet coefficients. Trans. Signal Process. 41(12), 3445–3462 (1993). doi:10.1109/78.258085. URL http://dx.doi.org/10.1109/78.258085

114. Skodras, A., Christopoulos, C., Ebrahimi, T.: The JPEG 2000 still image compression standard. IEEE Signal Process. Mag. 18, 36–58 (2001)

115. Slepian, D., Wolf, J.: Noiseless coding of correlated information sources. IEEE Trans. Inf. Theory 19(4), 471–480 (2006). doi:10.1109/TIT.1973.1055037. URL http://dx.doi.org/10.1109/TIT.1973.1055037

116. Stitt, G., Vahid, F., Nematbakhsh, S.: Energy savings and speedups from partitioning critical software loops to hardware in embedded systems. ACM Trans. Embed. Comput. Syst. 3(1), 218–232 (2004). doi:10.1145/972627.972637. URL http://doi.acm.org/10.1145/972627.972637

117. Stojanovic, M.: Underwater acoustic communications: Design considerations on the physical layer. In: Fifth Annual Conference on Wireless on Demand Network Systems and Services, 2008 (WONS 2008). pp. 1–10 (2008). doi:10.1109/WONS.2008.4459349

118. Svensson, H.: Reconfigurable architectures for embedded systems. Ph.D. thesis, Lund University (2008)

119. Szeliski, R.: Image alignment and stitching: A tutorial. Found. Trends Comput. Graph. Vis. 2(1), 1–104 (2006). doi:10.1561/0600000009. URL http://dx.doi.org/10.1561/0600000009

120. Taubman, D.: High performance scalable image compression with ebcot. Trans. Image Process. 9(7), 1158–1170 (2000). doi:10.1109/83.847830. URL http://dx.doi.org/10.1109/83.847830

121. Telle, N., Luk, W., Cheung, R.: Customising hardware designs for elliptic curve cryptography. In: Pimentel, A., Vassiliadis, S. (eds.) Computer Systems: Architectures, Modeling, and Simulation. Lecture Notes in Computer Science, vol. 3133, pp. 274–283. Springer, Berlin (2004). doi:10.1007/978-3-540-27776-7_29. URL http://dx.doi.org/10.1007/978-3-540-27776-7_29

122. The International Society of Automation: ISA100, Wireless Systems for Automation. www.isa.org/isa100 (2008)

123. Todman, T., Constantinides, G., Wilton, S., Mencer, O., Luk, W., Cheung, P.: Reconfigurable computing: Architectures and design methods. IEE Proc. Comput. Digit. Tech. 152(2), 193–207 (2005). doi:10.1049/ip-cdt:20045086

124. Treisman, A., Gelade, G.: A feature-integration theory of attention. Cogn. Psychol. 12(1), 97–136 (1980)

125. Tsapatsoulis, N., Rapantzikos, K.: Wavelet based estimation of saliency maps in visual attention algorithms. In: Proceedings of the 16th International Conference on Artificial Neural Networks - Volume Part II (ICANN'06), pp. 538–547. Springer, Berlin (2006). doi:10.1007/11840930_56. URL http://dx.doi.org/10.1007/11840930_56

126. Tsapatsoulis, N., Rapantzikos, K., Pattichis, C.S.: An embedded saliency map estimator scheme: Application to video encoding. Int. J. Neural Syst. 17(4), 289–304 (2007). URL http://dblp.uni-trier.de/db/journals/ijns/ijns17.html#TsapatsoulisRP07

127. Tsekoura, I., Selimis, G., Hulzink, J., Catthoor, F., Huisken, J., de Groot, H., Goutis, C.: Exploration of cryptographic ASIP designs for wireless sensor nodes. In: 2010 17th IEEE International Conference on Electronics, Circuits, and Systems (ICECS), pp. 827–830 (2010). doi:10.1109/ICECS.2010.5724640

128. Tyson, Romas, A., Siti Intan, P., Adiono, T.: A pipelined double-issue mips based processor architecture. In: International Symposium on Intelligent Signal Processing and Communication Systems, 2009 (ISPACS 2009), pp. 583–586 (2009). doi:10.1109/ISPACS.2009.5383771

129. Urban, F., Follet, B., Chamaret, C., Le Meur, O., Baccino, T.: Medium spatial frequencies, a strong predictor of salience. Cogn. Comput. **3**, 37–47 (2011). doi: 10.1007/s12559-010-9086-8. URL http://hal.inria.fr/inria-00628096

130. Viola, P., Jones, M.J.: Robust real-time face detection. Int. J. Comput. Vis. **57**(2), 137–154 (2004). doi:10.1023/B:VISI.0000013087.49260.fb. URL http://dx.doi.org/10.1023/B:VISI.0000013087.49260.fb

131. Wagner, R., Nowak, R., Baraniuk, R.: Distributed image compression for sensor networks using correspondence analysis and super-resolution. In: Proceedings of the 2003 International Conference on Image Processing, 2003 (ICIP 2003), vol. 1, pp. I–597–600 (2003). doi: 10.1109/ICIP.2003.1247032

132. Wallace, G.K.: The jpeg still picture compression standard. Commun. ACM **34**(4), 30–44 (1991). doi:10.1145/103085.103089. URL http://doi.acm.org/10.1145/103085.103089

133. Walther, D., Koch, C.: 2006 special issue: Modeling attention to salient proto-objects. Neural Netw. **19**, 1395–1407 (2006). doi:10.1016/j.neunet.2006.10.001. URL http://dl.acm.org/citation.cfm?id=1219169.1219421

134. Warneke, B., Scott, M., Leibowitz, B., Zhou, L., Bellew, C., Chediak, J., Kahn, J., Boser, B., Pister, K.: An autonomous 16 mm3 solar-powered node for distributed wireless sensor networks. In: Proceedings of IEEE Sensors, 2002, vol. 2, pp. 1510–1515 (2002). doi:10.1109/ICSENS.2002.1037346

135. Wheeler, F., Pearlman, W.: Spiht image compression without lists. In: Proceedings of the 2000 IEEE International Conference on Acoustics, Speech, and Signal Processing, 2000, vol. 6 (ICASSP '00), vol.4, pp. 2047–2050 (2000). doi:10.1109/ICASSP.2000.859236

136. Winbond: Winbond W968D6DKA product page. http://www.winbond.com/hq/enu/ProductAndSales/ProductLines/MobileRAM/PseudoSRAM/W968D6DKA.htm (2008)

137. Wu, M., Chen, C.W.: Collaborative image coding and transmission over wireless sensor networks. EURASIP J. Appl. Signal Process. **2007**(1), 223–223 (2007). doi: 10.1155/2007/70481. URL http://dx.doi.org/10.1155/2007/70481

138. Wyner, A.D., Ziv, J.: The rate-distortion function for source coding with side information at the decoder. IEEE Trans. Inf. Theory **22**, 1–10 (1976)

139. Xilinx: MicroBlaze Soft Processor Core. http://www.xilinx.com/tools/microblaze.htm (2012)

140. Xilinx: Spartan-3 datasheet. http://www.xilinx.com/support/documentation/data_sheets/ds099.pdf (2012)

141. Xilinx: Xilinx Artix-7. http://www.xilinx.com/products/silicon-devices/fpga/artix-7/index.htm (2012)

142. Xilinx: Xilinx XPower. http://www.xilinx.com/products/design_tools/logic_design/verification/xpower.htm (2012). Accessed Nov 2012

143. Zeidman, B.: Introduction to Verilog. Swiss Creek Publications (2000)

144. Zeidman, B.: Designing With FPGAs and CPLDs, 1st edn. CRC Press, Boca Raton (2002)

145. Zhang, J., Orlik, P., Sahinoglu, Z., Molisch, A., Kinney, P.: Uwb systems for wireless sensor networks. Proc. IEEE **97**(2), 313–331 (2009). doi:10.1109/JPROC.2008.2008786

146. ZigBee Alliance: ZigBee Standards. http://www.zigbee.org/Standards/Downloads.aspx (2007)

147. Zulkifli, M., Yudhanto, Y., Soetharyo, N., Adiono, T.: Reduced stall mips architecture using pre-fetching accelerator. In: International Conference on Electrical Engineering and Informatics, 2009 (ICEEI '09), vol. 02, pp. 611–616 (2009). doi:10.1109/ICEEI.2009.5254742

148. Zvikhachevskaya, A., Markarian, G., Mihaylova, L.: Quality of service consideration for the wireless telemedicine and e-health services. In: IEEE Wireless Communications and Networking Conference, 2009 (WCNC 2009), pp. 1–6 (2009). doi:10.1109/WCNC.2009.4917925

Index

L. Ang et al., *Wireless Multimedia Sensor Networks on Reconfigurable Hardware*,
DOI 10.1007/978-3-642-38203-1, © Springer-Verlag Berlin Heidelberg 2013

Printed in the United States
by Bookmasters

Printed in the United States
By Bookmasters